Practical Low-Cost IC Projects

by

Herbert Friedman

A Revision of

30 IC Projects

By Herbert Friedman

Howard W. Sams & Co., Inc.
4300 WEST 62ND ST. INDIANAPOLIS, INDIANA 46268 USA

SECOND EDITION
FIRST PRINTING—1979

International Standard Book Number: 0-672-21599-3
Library of Congress Catalog Card Number: 79-62996

Printed in the United States of America.

PREFACE

There are two kinds of electronics hobbyists: the *experimenter,* who builds circuits primarily to find out how things work, and the *builder,* who puts together bits and pieces of individual circuits and comes up with some project that will be put to immediate practical use.

The projects in this book are intended primarily for the *builder* who wants to improve or expand some other area of his electronics interest, such as tape recording, high fidelity, photography, BCB DX'ing, even home security. Naturally, there are also projects of particular interest to the active radio amateur and CB'er.

Although most of the projects can stand alone as individual accessories, wherever possible the circuits have been designed so that they can be easily ganged with one or more other projects. For example, the 4-channel *microphone mixer* can drive the 10×/100× *instrument amplifier* to produce a complete mixer with a line level output. Similarly, *the low-pass filter* can be combined with *the high-pass filter* for frequency control at both ends of the audio range. The audiophile might consider building one single test instrument from the *af oscillator* and *stereo balancer,* while the photo hobbyist will find that the *light meter* and *enlarger timer* assembled in a single cabinet will give the equal of similar commercially made accessories, but at a considerably reduced cost.

Because integrated-circuit projects utilize a wide range of power-supply voltages, we have included several different regulated and unregulated power-supply projects so that the hobbyist doesn't run up an expensive bill for replacement batteries. The power-supply requirement for each circuit has been tailored to a "standard" voltage value that is easily attained by using generally available batteries or power-supply transformers. These voltages, in most instances, will be similar to those for projects published in hobbyist magazines, so a power-supply project from this book can be used for many diverse projects from other sources.

Because we have attempted to make each circuit universal in the sense that it will easily connect to other circuits and work well with a relatively broad range of power-supply voltages, some do not deliver all their inherent capacity in terms of amplification, sensitivity, etc. Each circuit will do what is claimed, but the builder might find that a change in power-supply voltage or resistance value will produce a more desirable level of performance. Changing the circuit, however, might make it more difficult to gang two or more circuits. But if an optimized circuit is preferred, there's no harm in making changes. Circuits that might prove critical and tricky to adjust have been designed with "protection" values; other than a direct short circuit or an extreme power-supply voltage change, no damage will be done to the integrated circuit if the performance is optimized.

Because there is a wide latitude in circuit modifications, most circuits will be of value to the *experimenter* who wants to see how things work, even though each project's primary value is for the *builder* who desires a functional item of equipment as the end result of his efforts.

HERBERT FRIEDMAN

Contents

FILTERS AND OSCILLATORS

HOBBY CIRCUITS

Introduction

The modern integrated circuit—or *IC* as it is generally termed—serves but one function: to reduce the many discrete components needed for a complex electronic circuit into a single device. Through specialized photographic techniques the modern IC can contain the equivalent of hundreds of transistors, resistors, capacitors, and even inductors, and can sell for less than ten dollars, or as little as fifty cents on the "surplus" market.

In many instances it is possible to utilize individual circuit elements of a single IC as individual transistors or as multistage amplifiers; but then we are back to discrete component assembly and the IC serves as just another solid-state device which could easily be replaced by one or two transistors. In this instance we wind up with no savings in cost or assembly time; we are not making the best possible use of an IC.

The primary purpose of an IC project is to reduce the number of necessary components by the largest possible factor: 10X, 100X, or whatever. Total construction costs and assembly time should be similarly reduced to have a justification for using ICs. All the projects in this book have been selected on this basis; you could not duplicate any project using discrete components for twice, three times, five times, and, in some instances, ten times what it costs to build with ICs. As far as ease of assembly is concerned, most of the projects in this book can be built in a few hours, yet it would take several nights' effort to obtain the same functions and performance with discrete components.

Most of the projects are "basic building blocks" which you can "drop into" other more complex projects. For example, the timed auto burglar alarm can also be used as an alarm bell cut-off for an existing home or apartment burglar alarm. The tape-head preamplifier, which can be built for less than $1, will turn a "surplus" $15 8-track cartridge mechanism into a $75 tape player. The instrument amplifier will give the sensitivity of a lab-grade oscilloscope to a low-cost service-grade oscilloscope. In short, the projects in this book are intended for the average *electronics experimenter* but are not limited to the experimenter only. You'll find projects for the audio enthusiast, stereophile, tape fan, radio amateur, and CB'er.

An important consideration for each project is the general availability of components. In most instances it has been possible to use ICs readily available from "surplus" parts distributors at rock bottom prices. For example, the 741 operational amplifier is used in many projects because it is almost universally available at low cost; the 741 is also essentially goof-proof. Replacing the specified 741 with a "precision operational amplifier" will not make the project(s) work any better, it will only serve to increase the overall costs. In other instances we have used ICs which, we have been told, will soon be available as "surplus." In a few instances ICs are used which are still available only from "industrial parts distributors" because the projects won't work well with substitutes. In each case, we have specified ICs from nationally distributed "brand names" generally available from local distributors.

If you have built any projects recently, you know that "transistor" capacitors often represent the greatest expense, with the total cost of the capacitors generally exceeding the cost of all other components except the cabinet or housing. With very few exceptions, every project in this book will work with the lowest-quality capacitors—whatever you can get at the lowest possible price from any source. Do not substitute higher-grade capacitors than are specified; they will not improve overall performance but just increase the total costs. Similarly, resistors should be ½ watt, 10% unless specified otherwise. Only in microphone and tape-head preamplifiers are 1% or low-noise resistors justified, and even here the common resistor will work well.

Construction details are provided when necessary, such as "use a heat sink." If a heat sink is not called for—even in a "power project"—there's no need to use one. If there are no specific instructions, the circuit can be assembled in any manner in any cabinet (wood or plastic, or no cabinet), provided reasonable care is taken in assembly.

To make things as easy as possible, two types of capacitor symbols are used. An electrolytic-type capacitor will have a "+" over one line and must be installed with the correct polarity. If there is no "+" symbol indicated, the capacitor can be installed in any manner.

As a general rule, the voltage rating of a capacitor should be at least equal to, or higher than, the power-supply voltage, unless specified otherwise.

Potentiometers can be any taper unless a particular taper is specified, such as "audio" or "linear." When batteries are specified, do not use a smaller size than recommended; current requirements for a project are taken into account for the battery type suggested.

Do not fear to tackle any project in this book. All have been geared for the typical experimenter, and it does not matter whether you are a beginner or an experienced builder. Your success is assured; your reward will be a useful project, enjoyment, and deep personal satisfaction.

Power Supplies

1. Bipolar Power Supply

Many ICs use the *operational amplifier* as the basic circuit element. These amplifiers usually require a bipolar power source, meaning both a positive and a negative voltage with respect to "ground." In many instances, and certainly for the projects in this book, a regulated power supply is not needed as the circuit will work over a reasonable range of power-supply voltage, say ±9 volts to ±15 volts.

Power transformer T1 can be any center-tapped filament transformer, such as 12.6 VCT or 24 VCT; alternatively, T1 can be a multitapped (multivoltage) "experimenter" or "surplus" power transformer such as sold by many mail order distributors.

The rectifier consisting of silicon diodes SR1 through SR4 can be replaced by a single "bridge rectifier" with appropriate ratings. Bridge rectifiers are often available from surplus distributors for far less than the cost of four silicon diodes.

Note, carefully, that there are three output connections: ground, +E, and −E. When connecting to a bipolar-powered project, make certain the ground wire is connected to the project.

The maximum current that can be attained is limited by the rectifier and/or transformer ratings. One ampere is generally sufficient for most hobbyist applications, although there's no need for more current than the project requires. If the project uses, say, 50 milliamperes, transformer T1 can be rated 50 mA, though we suggest 1-A diodes be used for protection against low-resistance "shorts."

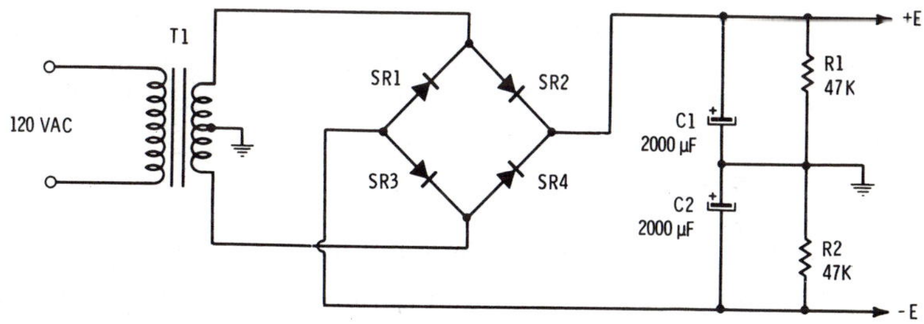

DESIRED OUTPUT VOLTAGE	TRANSFORMER SECONDARY VOLTAGE	ACTUAL DC
±9 VDC	12.6 V CENTER TAPPED	± 8 V
±9 VDC	14 V CENTER TAPPED	± 9 V
±12 VDC	18 V CENTER TAPPED	± 12.5 V
±15 VDC	22 V CENTER TAPPED	± 15 V
±16 VDC	24 V CENTER TAPPED	± 16.5 V

Bipolar power supply.

PARTS LIST

Item	Description
T1	Power transformer (see text)
SR1 to SR4	Silicon diode, 50 PIV, 1 A
C1, C2	2000-μF, 25-volt electrolytic capacitor
R1, R2	47K, ½-watt resistor

2. Regulated 5 V DC/3 A Power Supply

Five volts dc is almost a "universal" power source for digital IC projects. This power supply, which provides up to 3 amperes of current, uses a simple 3-terminal regulator requiring no external components, yet it provides a "tight" lab-grade regulation.

Either a metal chassis or a separate heat sink should be used for IC1. Note that the metal case of the IC is connected to ground, so an insulated mounting socket is unnecessary.

Transformer T1 can be a 24-VCT or 25.5-VCT filament type, rated at least 3 A.

Capacitor C2 should be connected as close as possible to the IC.

The regulator can be National type LM223 or LM323, whichever is available.

(5 V. REGULATORS)

309H — 200mA. MAX OUTPUT CURRENT

309K — 1 AMP. MAX. OUTPUT CURRENT

(123K
223K)

323K — 3 AMP. MAX. " "

7805 T — 1 AMP. MAX. " "

7806 K R.P 2.5 AMP MAX. " "

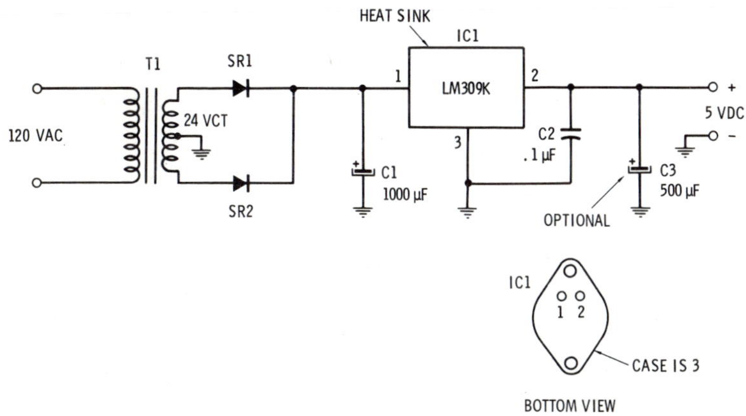

Regulated power supply.

PARTS LIST

Item	Description
C1	1000-μF, 20-volt, or higher, electrolytic capacitor
C2	0.1-μF, 5-volt, or higher, capacitor
C3	500-μF, 5-volt, or higher, electrolytic capacitor (optional)
IC1	National LM309K 5-volt dc regulator
SR1	Silicon diode, 25 PIV, 3 A or higher
T1	Transformer (see text)

3. Preset Voltage Regulator

This low-cost easy-to-build voltage regulator will provide any selected voltage from 5 V dc to 28 V dc at currents up to 200 mA. The precise voltage output is set by the ratio of R1 and R2. The chart shows the total resistance value (R1 + R2) for the voltage ranges.

Note the power transformer has no center tap. Bridge rectifier BR1 can be four individual silicon diodes as shown, rated at least 50 PIV at 0.5 A. Transformer T1 secondary voltage must provide a dc voltage output from the rectifier of at least 3 volts higher than the final desired output voltage. T1 secondary output voltage (ac) is 0.707 × V_{in} (V_{in} being the output of the bridge rectifier). For example, if V_{in} is to be 18 volts, T1 secondary voltage should be 12.6 volts.

Voltage V_{out} is equal to V_{in} × R1 + R2. The resistance values shown in the chart are not critical; you can trim the value of R1 to attain the precise desired output voltage.

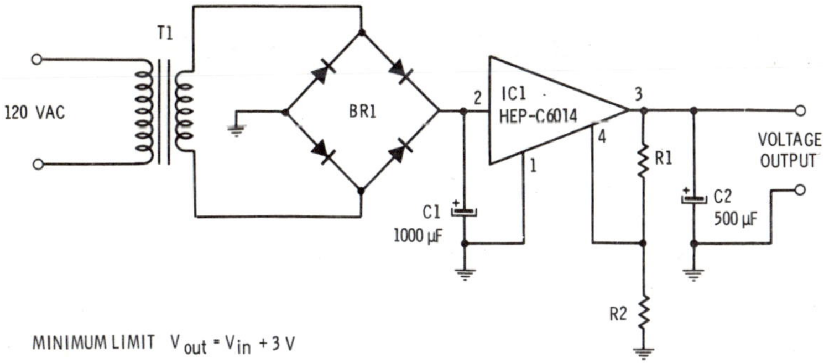

MINIMUM LIMIT $V_{out} = V_{in} + 3\text{ V}$

$$V_{out} = \frac{V_{in} \times R1}{R1 + R2}$$

IC1
TOP VIEW

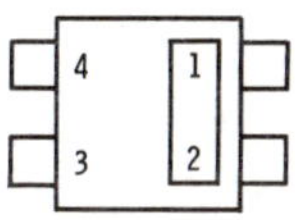

R1 + R2		
=	2400 Ω	5 TO 10 V
=	4700 Ω	11 TO 20 V
=	6800 Ω	21 TO 28 V

MAXIMUM RESISTANCE VALUES

Preset voltage regulator.

PARTS LIST

Item	Description
C1	1000-μF electrolytic capacitor, rated at least V_{in}
C2	500-μF electrolytic capacitor, rated at least V_{out}
IC1	Motorola HEP-C6014
BR1	Bridge rectifier or silicon diodes rated 50 PIV, 0.5 A
R1, R2	See text and chart
T1	See text

4. Walkie-Talkie Power Pack

Most of the high-power (3 watt, 5 watt) Citizens band walkie-talkies and quite a few 2-meter fm amateur radio hand-held transceivers make excellent base-station performers when connected to a rooftop antenna. But continuous use of the battery-operated units can produce astronomical power costs; you can easily use two or three sets of batteries a day.

If your transceiver or walkie-talkie can be connected through an optional power cable to an automobile battery (usually through the cigarette lighter socket), or if the unit has provision for an external heavy-duty battery pack, you can substitute this regulated ac-to-dc converter, which provides 14.5 volts of output at a continuous 1 ampere.

Transformer T1 can be any transformer with a secondary output of 14 to 16 volts ac and with a current capacity of at least 1 ampere. The four silicon rectifiers comprising diode bridge rectifier BR1 should be rated at least 50 PIV 1 A. A bridge rectifier with this rating is often available at a lower cost than four diodes on the "surplus" market. The regulator, IC1, is available in two styles as shown; depending on the style, either the case or the tab is grounded. Note that if you use the IC "T" (tab) package, the chamfered edge must face up if the leads are to conform to the terminal arrangement shown. You must use a heat sink; either construct the power supply on a small aluminum chassis, or use a small power transistor heat sink (such as those available in Motorola's HEP line).

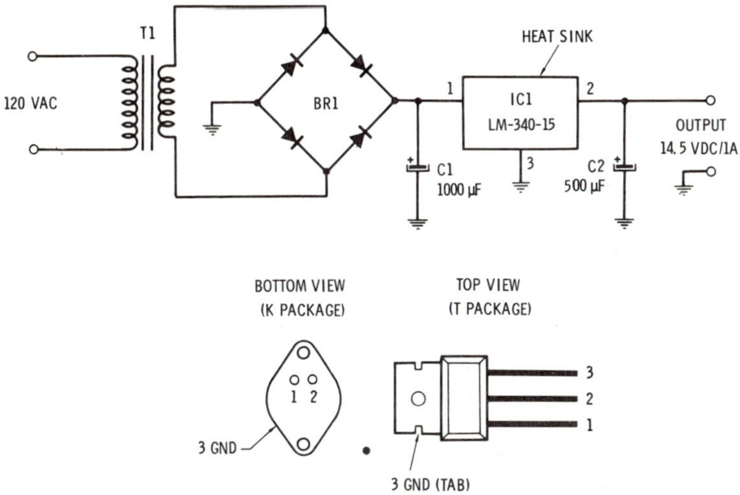

Walkie-talkie power-pak.

PARTS LIST

Item	Description
C1	1000-μF or 2000-μF electrolytic capacitor, rated 25 V dc or higher
C2	500-μF electrolytic capacitor, rated 15 V dc or higher
BR1	Bridge rectifier, 50 PIV, 1 A (see text)
IC1	National LM-340-15 voltage regulator
T1	Power transformer (see text)

Amplifiers and Boosters

5. A-M Booster

Is your favorite a-m or shortwave station usually buried in the noise? If there's enough signal strength for you to hear the full program through the noise level, this booster will give you more than enough extra sensitivity to pull the signal out into the clear. Depending on your particular radio or receiver, the booster delivers from 10 dB to 15 dB of additional "sensitivity"—up to 3 S-units more signal. The frequency coverage of the booster is from the a-m broadcast band to about 27 MHz.

For a-m radios with a built-in "loop" antenna, the output of IC1 should be connected to any type of "Loopstick" antenna coil. Connect a "long-wire" antenna to binding post BP1 and place Loopstick L1 adjacent and parallel to the radio antenna coil. Then, just tune in the desired station. (The a-m booster is a broad-band device and requires no tuning of its own.) If you get feedback in the radio, make certain L1 is positioned as far as possible from BP1. The least feedback potential exists when the booster is constructed in a metal enclosure, though plastic can be used if the inputs and output are kept well apart. You can connect L1 to the circuit through miniature coaxial cable, such as RG-59U, up to three feet in length.

The booster can also be used with shortwave receivers that have an "antenna trimmer." Use the alternate output circuit shown (with phono jack), and connect the booster to the receiver with up to three feet of miniature coaxial cable. After the station is tuned in, adjust the antenna trimmer for maximum strength.

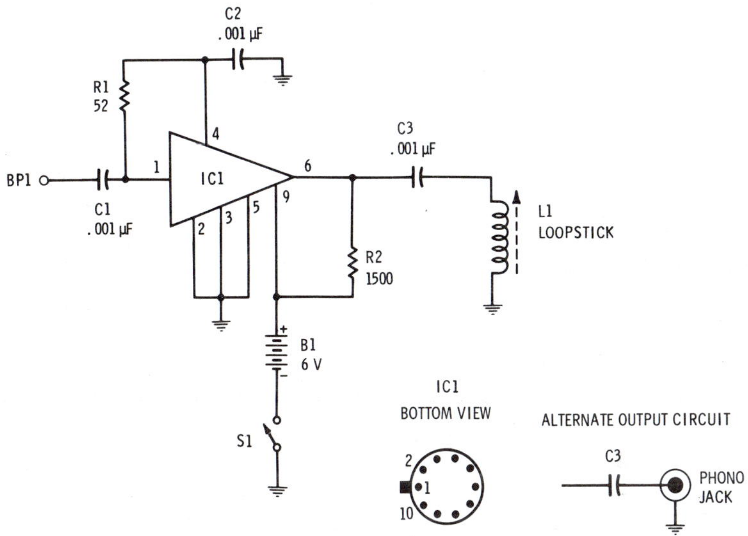

A-m booster.

PARTS LIST

Item	Description
R1	52-ohm, ½-watt resistor
R2	1500-ohm, ½-watt resistor
C1, C2, C3	0.001-μF, 500-V dc ceramic disc or Mylar capacitor
IC1	Motorola HEP 590 integrated circuit
L1	Loopstick antenna
BP1	Insulated binding post
B1	6-V battery, Burgess Z4 or equiv
S1	Switch, spst

6. 10×/100× Instrument Amplifier

When signals get so weak they barely move the pointer of your vtvm, or when they make no more than a squiggle on your scope, just connect this instrument amplifier between the unknown voltage and your test equipment and you will get full-scale or full-screen readings.

The input impedance of the amplifier is 1 megohm, and the frequency response extends from dc to beyond 20,000 Hz. If you do not require this much bandwidth, you can substitute a 741 for IC1 that gives a bandwidth of dc to about 10,000 Hz. Capacitor C5 is eliminated if you use a 741. S1 selects the gain: 100× when S1 is open; 10× when S1 is closed.

Reasonable care must be taken during assembly to prevent high-frequency self-oscillation. Capacitors C2 and C3 should be installed as closely as possible to the IC pin terminals. The input circuit connected to pin 3 should be reasonably spaced from the output components connected to pin 6.

After the amplifier is completed, connect a vtvm across the output, from pin 6 to ground, and adjust R4 for a zero dc meter reading.

The amplifier should be installed in a metal cabinet to prevent hum pickup. Drill a hole in the cabinet opposite R4 so that the DC output drift can be adjusted from time to time. If you are concerned primarily with ac measurements, the dc zero drift is unimportant, a 0.25-μF capacitor rated 25 vdc (or higher) can be connected in series with the output of the amplifier.

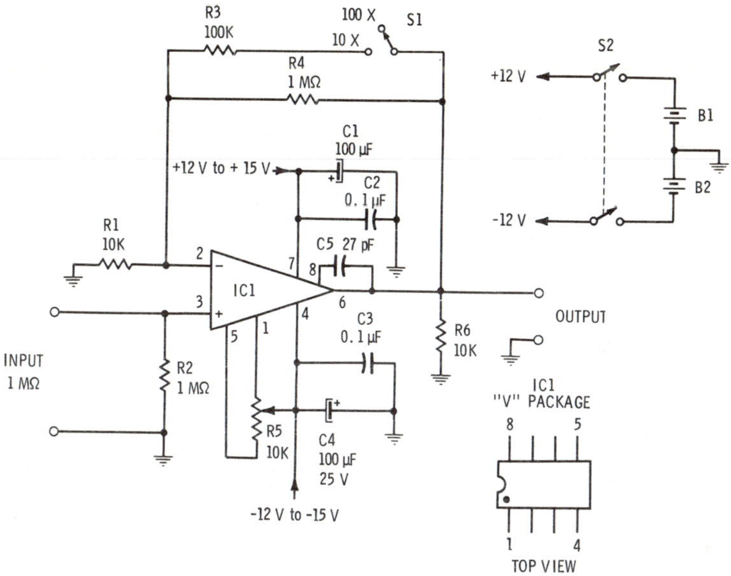

A 10×/100× instrument amplifier.

PARTS LIST

Item	Description
R1, R6	10,000-ohm, ½-watt resistor
R2, R4	1-megohm, ½-watt resistor
R3	100,000-ohm, ½-watt resistor
R5	10,000-ohm miniature or trimmer potentiometer
C1, C4	100-µF, 15-V dc, or higher, electrolytic capacitor
C2, C3	0.1-µF, 25-V dc, or higher, Mylar capacitor
C5	27-pF ceramic disc, 25-V dc, or higher
IC1	Operational amplifier, SE or NE531, or 741 (see text)
S1	Switch, spst
S2	Switch, dpst
B1, B2	Battery within range of 12 to 15 volts. Eveready 304116 smoke-alarm battery is recommended.

7. Budget Tape-Head Preamplifier

From time to time surplus tape, cartridge, and cassette mechanisms are available at rock bottom prices, often less than $20. These mechanisms are complete except for the electronics. This tape-head preamplifier can be built from surplus parts for well under $4 and will allow the output of the tape player to be fed into any high-level amplifier or preamplifier.

Ordinary transistor radio-type 9-volt batteries will provide many hours of service because their current drain is minimal, just a fraction of normal transistor-radio current requirements. Take particular note that the "ground," or "shield," wire from the tape head is not connected to the circuit ground, but to the junction of R1 and R2.

If used in a motor vehicle, a 12-volt auto battery can be substituted for B1 and B2. Since this will result in slightly less overall gain, it might be necessary to provide a simple rf filter between the battery connection and the preamplifier.

Capacitor C5 can be a 0.1-μF unit if the preamplifier works into a load of 50,000 ohms or higher, such as a tube-type amplifier or the auxiliary input of a high-fidelity amplifier. If the preamplifier will be connected to a solid-state amplifier of low or unknown input impedance, a 2-μF capacitor should be used.

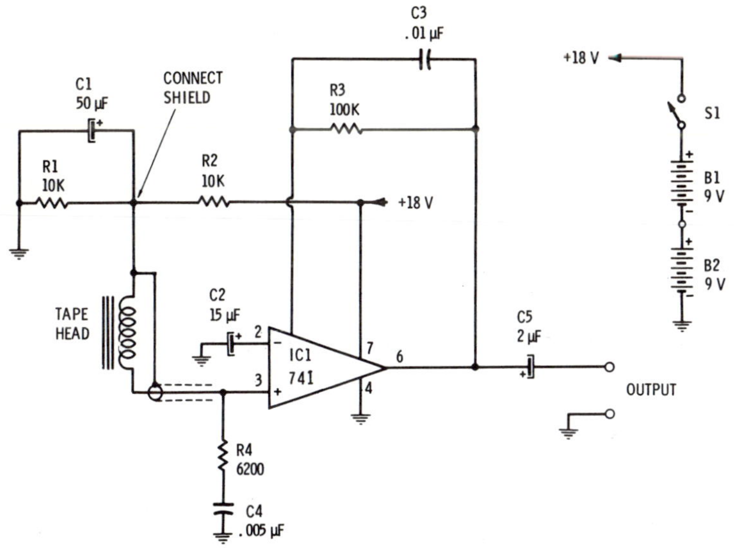

Budget tape-head preamplifier.

PARTS LIST

Item	Description
R1, R2	10,000-ohm, ½-watt resistor
R3	100,000-ohm, ½-watt resistor
R4	6200-ohm, ½-watt resistor
C1	50-µF, 25-V dc electrolytic capacitor
C2	15-µF, 25-V dc electrolytic capacitor
C3	0.01-µF, 25-V dc, or higher, capacitor
C4	0.005-µF, 25-V dc, or higher, capacitor
C5	2-µF, 15-V dc, or higher, electrolytic capacitor (see text)
IC1	Type 741 operational amplifier
B1, B2	9-volt transistor-radio-type battery (see text)
S1	Switch, spst

8. General-Purpose Audio Amplifier

This gem of an amplifier costs very little to build, yet it can serve as a microphone preamplifier, high-impedance to low-impedance converter, a line amplifier or a Citizens band talk power booster. As a microphone preamplifier it will handle any type of microphone from a 50-ohm low-impedance dynamic type to a high-impedance ceramic or crystal type. The input impedance is well in excess of 1 megohm, while the voltage gain is 40 dB.

The circuit shown in A requires a bipolar power supply. The amplifier load can be 1000 ohms or higher. The modified circuit shown in B will work from a single-ended power supply (one voltage against ground), but it requires resistors R4 and R5, and capacitor C3. When C3 is 0.22 μF, the load must be greater than 50,000 ohms for a "flat" (20-20,000 Hz) frequency response. If C3 is 5 μF (rated 25 V dc or higher), the load may be as low as 1000 ohms. With C3 as 5 μF, the response will be down 3 dB at 20 Hz with a 1000-ohm load. As the load impedance is increased above 1000 ohms, the attenuation at 20 Hz is reduced. For an electrolytic-type C3 (1 μF or larger), the positive terminal must connect to pin 6 of the IC.

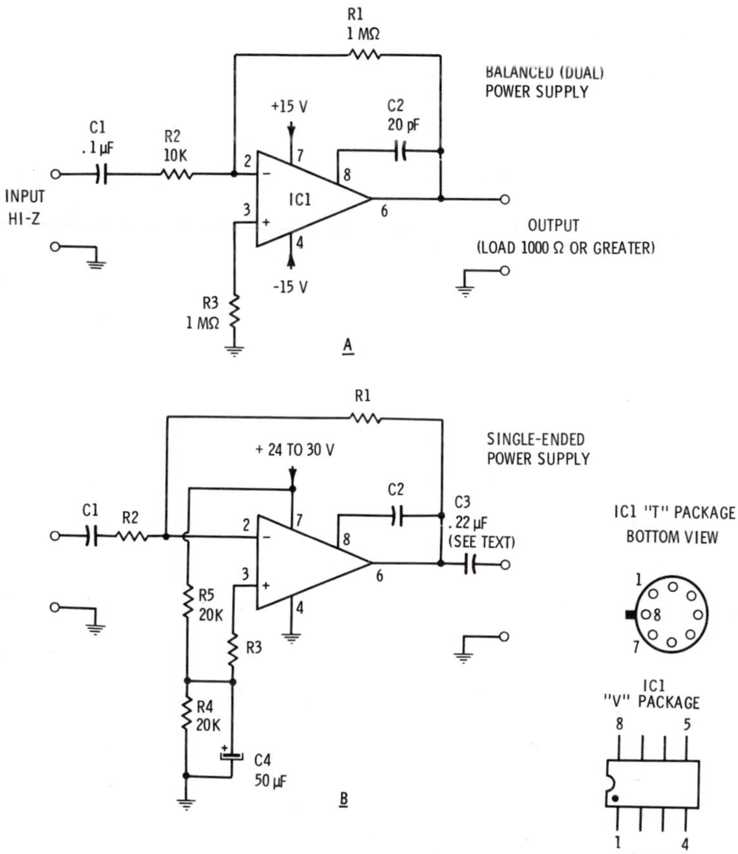

General-purpose audio amplifier.

PARTS LIST

Item	Description
R1	1-megohm, ½-watt resistor
R2	10K, ½-watt resistor (low-noise type suggested)
R3	1-megohm, ½-watt resistor (low-noise type suggested)
R4, R5	20K, ½-watt resistor
C1	0.01-µF, 500-V dc ceramic disc capacitor
C2	20-pF, 100-V dc ceramic disc capacitor
C3	0.22-µF, 25-V dc capacitor (see text)
C4	50-µF, 50-V dc, or higher, electrolytic capacitor
IC1	Signetics NE or SE 531 operational amplifier
Miscellaneous	Metal enclosure or shield

9. High-Gain Microphone Preamplifier

One problem with using unbalanced (single ended) microphone lines is hum and noise pickup when the cable length approaches 25 feet. Use a balanced microphone connection, which is generally possible with low-impedance microphones, and you can run cable length into the hundreds of feet without noise pickup. The secret behind the noise-free pickup is cancellation at the preamplifier. The noise is still induced in the lines, but since it is equal in both lines it can be cancelled if the input stage is a "differential amplifier"—which cancels common-mode signals.

This balanced-amp provides 40-dB gain and can be used with balanced low-Z microphones rated 150 to 600 ohms. In the event you have a balanced 50-ohm microphone, which is somewhat unusual because nearly all 50-ohm "mikes" are unbalanced, you can use it with no modification to the circuit.

The current drain is very low and the circuit can be powered by standard transistor radio smoke-detector-type batteries in the 9- to 15-volt range. All resistors should be 1% for high common mode rejection, but if you have some 5% resistors, they can be used if you are willing to accept slightly less-than-normal noise cancellation.

Capacitors C2 and C3 should be installed as close as possible to the IC pins. Capacitors C1 and C4 are required for all power supplies, even batteries.

The IC pin connections will depend on the specific package. See other projects in this book for the various configurations for the SE and NE 531.

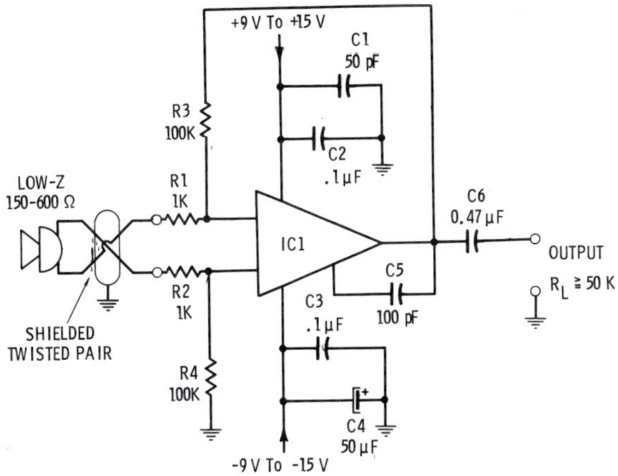

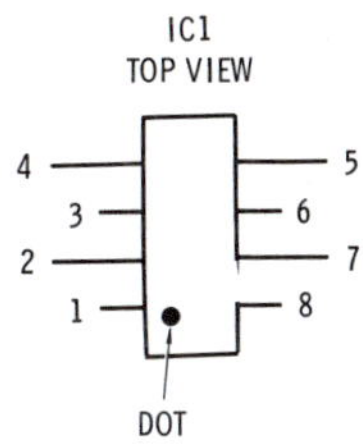

High-gain microphone preamplifier.

PARTS LIST

Item	Description
R1, R2	1000-ohm, ½-watt resistor (low noise 1% suggested)
R3, R4	100,000-ohm, ½-watt resistor (low noise 1% suggested)
C1, C4	50-µF, 25-V dc electrolytic capacitor
C2, C3	0.1-µF Mylar capacitor
C5	100-pF ceramic disc capacitor
IC1	Signetics NE or SE 531 operational amplifier

10. Bipolar Power Amplifier

With many modern audio circuits using operational amplifiers and bipolar power supplies, the easiest way to add a small low-cost monitoring amplifier is to stay in the bipolar power mode. Not only is the power-supply connection simplified, but the need for a large output-coupling capacitor is eliminated—a capacitor which can represent better than half the total cost.

This bipolar power amplifier will provide approximately 1-watt maximum output into a 16- to 22-ohm speaker load or into high-fidelity-type headphones. Although the power source is shown as ±9 V dc, it can be connected to ±12 V dc if a small clip-type heat sink is secured to the IC case. The overall voltage gain is approximately 36. If desired, the overall gain can be reduced 6 dB by opening the connection between pins 2 and 5.

The input impedance is 10,000 ohms. If a volume control is needed directly in front of the amplifier, use a 50,000-ohm audio-taper potentiometer.

The zero signal current is 15 mA, which precludes the use of standard transistor-radio-type batteries. If you must use a battery supply in preference to a line-powered supply, a battery such as the Eveready 276 is suggested.

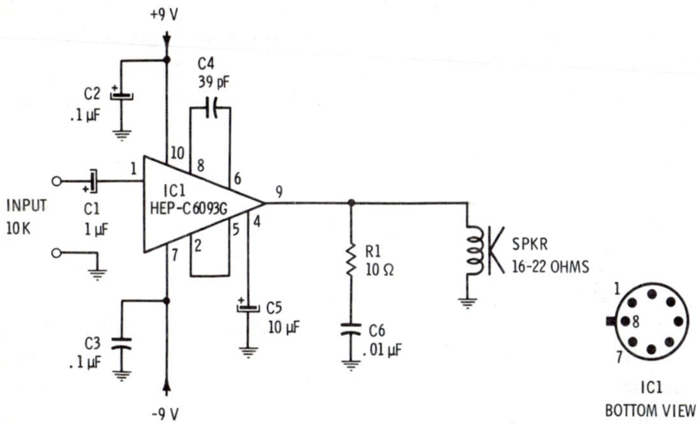

Bipolar power amplifier.

PARTS LIST

Item	Description
R1	10-ohm, ½-watt resistor
C1	1-μF, 10-V dc electrolytic capacitor
C2, C3	0.1-μF Mylar capacitor
C4	39-pF disc capacitor
C5	10-μF, 10-V dc electrolytic capacitor
C6	0.01-μF Mylar capacitor
IC1	Motorola HEP-C6093G
SPKR	16- or 20-ohm speaker (see text)
Miscellaneous	Volume control, batteries (see text)

11. Four-Channel Microphone Mixer

The best signal-to-noise ratio in a microphone mixer is always obtained when the volume control and/or mixer network are placed after the basic preamplification. In this instance a single IC provides four independent preamplifiers suitable for microphones in the 50- to 1000-ohm output-impedance range—the type of microphone normally provided, or used, with solid-state equipments.

Since the four preamplifiers are identical, for clarity the schematic shows the connections for only one circuit. The remaining circuits are identical. Note capacitor C3 and resistor R2. Even if you do not utilize one of the amplifiers, you must install C3 and R2 to avoid self-oscillation. It is quite possible that an active amplifier will work without C3 and R2. Try it. If there is no oscillation and complete stability, you save the cost of the components.

The power-supply requirements are 12 V dc at approximately 30 mA. Internal connections in the IC distribute the power to all the amplifier sections.

The outputs from all mixers are connected to the point(s) indicated as "A."

If you build a stereo model, use two separate ICs, one for each channel. To reduce cross talk between channels, increase the value of C5 to 200 μF. Since the IC is a high-frequency device, C6 must be used and should be installed as close as possible to the IC terminals.

PARTS LIST

Item	Description
R1	100-ohm, ½-watt resistor
R2	47-ohm, ½-watt resistor (see text)
R3	10,000-ohm audio-taper potentiometer
R4	20,000-ohm, ½-watt resistor
C1	0.1-μF Mylar or ceramic disc capacitor
C2	200-μF, 3-V dc electrolytic capacitor
C3	0.047-μF (0.05-μF) Mylar capacitor (see text)
C4	10-μF, 10-V dc electrolytic capacitor
C5	50-μF, 15-V dc electrolytic capacitor (see text)
C6	0.1-μF ceramic disc capacitor
IC1	RCA 3048 or Sylvania ECG-727

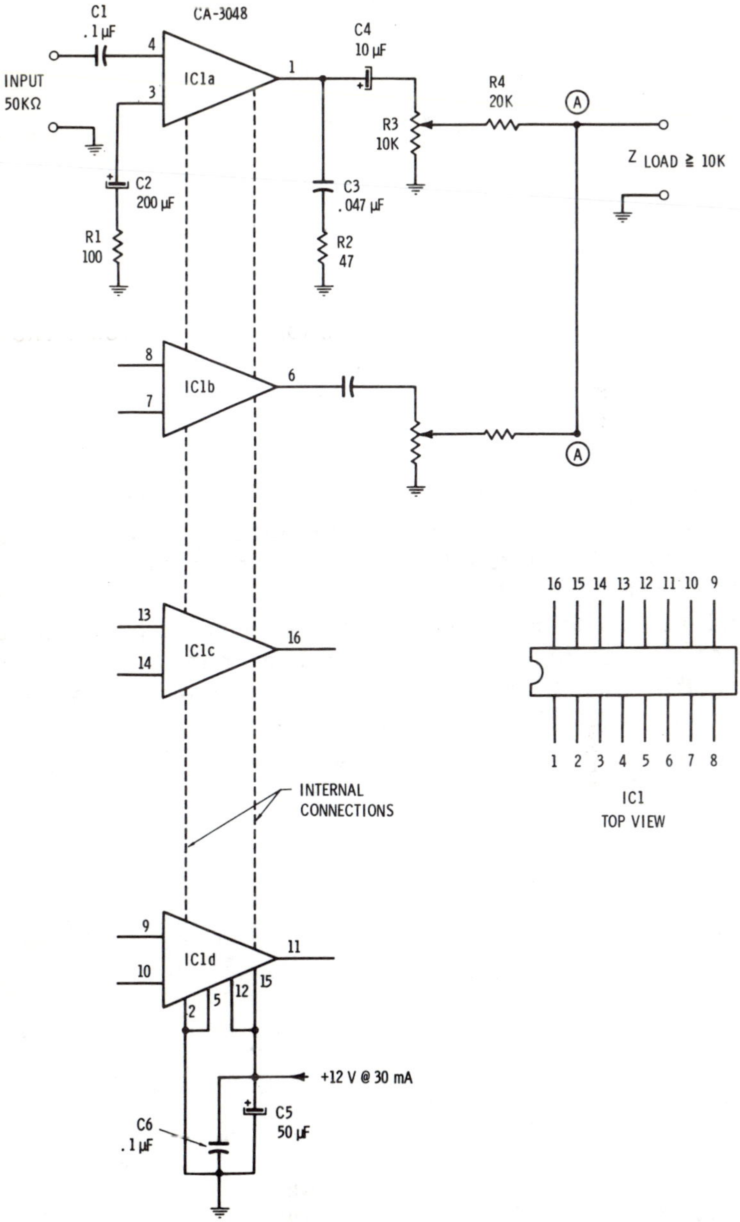

A 4-channel microphone mixer.

12. Remote-Control Audio Amplifier

As a general rule, the farther an amplifier volume control is located from the amplifier itself, the greater the potential for hum and noise pickup. But by using a dc voltage to control the amplifier gain, the problem of noise pickup is eliminated, and the volume control can be located hundreds, even thousands, of feet from the preamplifier and/or the signal source.

In this circuit the overall amplifier gain can be varied approximately 90 dB, while hum and noise picked up on the dc control wires to "volume control" R1 are filtered by C3 and are not impressed on the amplifier output signal.

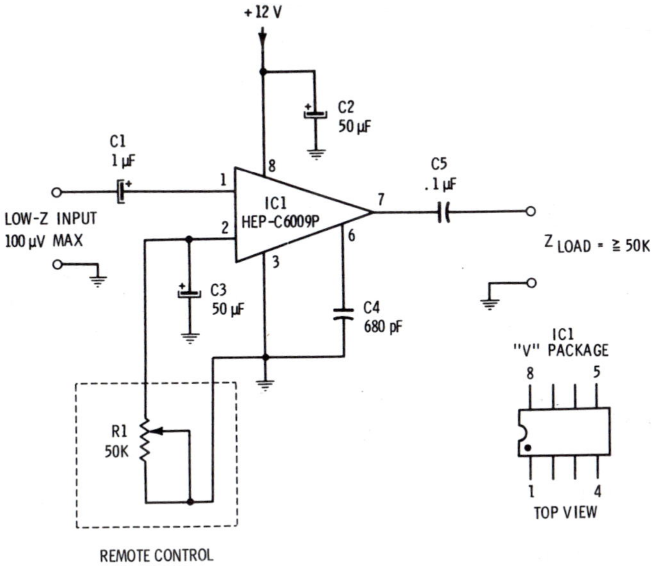

Remote-control audio amplifier.

PARTS LIST

Item	Description
R1	50,000-ohm potentiometer
C1	1-μF, 12-V dc electrolytic capacitor
C2	50-μF, 12-V dc electrolytic capacitor
C3	5-μF, 6-V dc electrolytic capacitor
C4	680-pF ceramic disc capacitor
C5	0.1-μF, Mylar or ceramic disc capacitor
IC1	Motorola HEP-C6009P electronic attenuator

13. Remote-Line Amplifier

It often happens that the recorder is on one end of the room and the microphone on the other—such as when a pianist attempts to record his own efforts. Or, although the public-address amplifier controls are upstairs in the control booth, it is desirable to have the volume control down in the audience. In both instances, the performer and/or control operator are a long way from the normal volume-level meters and controls. With this metered remote line amplifier, however, the performer or speaker can have full metered control of the volume level right at the microphone location.

The input signal can be a microphone or a line level. The output can be connected with optional circuit A to any high-impedance line-level input (recorder, amplifier, etc.). Optional circuit B, using a reasonably good-quality matching transformer —not a subminiature transistor type—is used for applications where it is necessary to drive a balanced line, such as for remote-broadcast applications.

Integrated circuit IC1 can be a type 741, or equivalent, if you can accept a slight rolloff from 10 kHz to 20 kHz. If a 741 is used, capacitor C2 is eliminated. For a flat 20 to 20 kHz frequency response, IC1 can be an NE/SE 531, in which case capacitor C2 must be used. The terminal arrangements will depend on the particular type of IC package. Both terminal arrangements—for the 741 and 531—can be found in other projects in this book.

Meter M1 is a standard miniature VU meter, such as those sold by Radio Shack and Lafayette Radio stores. It is necessary, however, to remove the internal rectifier of the meter so that only the meter movement and the scale remain. If you are not concerned with a precise meter reading, or if you do not require a VU scale, a 100-microampere meter can be substituted. The amplifier can be battery powered if desired, by using transistor-radio-type batteries.

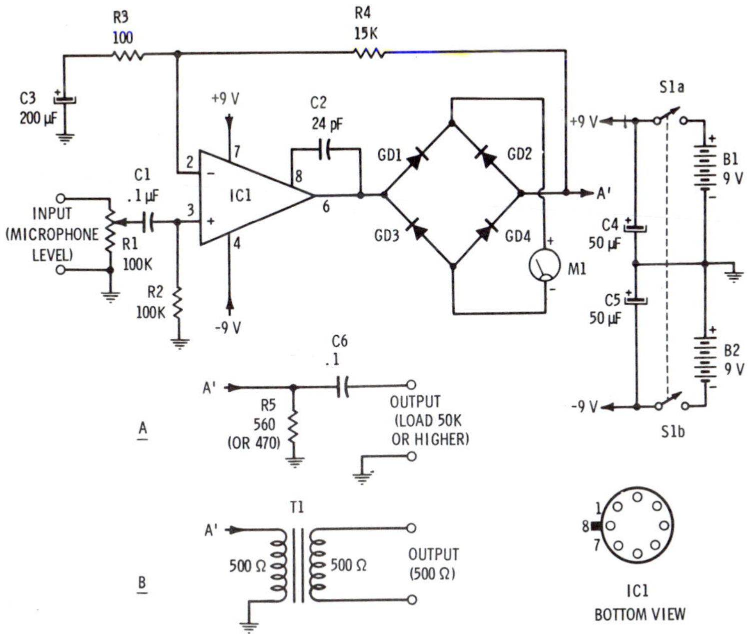

Remote line amplifier.

PARTS LIST

Item	Description
R1	100,000-ohm audio-taper potentiometer
R2	100,000-ohm, ½-watt resistor (low-noise type suggested)
R3	100-ohm, ½-watt resistor
R4	15,000-ohm, ½-watt resistor
R5	560-ohm, ½-watt resistor
C1, C6	0.1-µF Mylar capacitor
C2	20-pF disc capacitor (see text)
C3	200-µF, 10-V dc, or higher, electrolytic capacitor
C4, C5	50-µF, 10-V dc, or higher, electrolytic capacitor
IC1	See text
GD1, GD2, GD3, GD4	Germanium diode, 1N60 or equivalent
M1	VU meter (see text)
T1	500/500-ohm matching transformer (see text)
S1	Switch, dpst
B1, B2	9-volt battery, Burgess 2U6 or equivalent

14. Microphone Compressor

The point at which a public-address system starts to howl or at which a transmitter exceeds 100% modulation is determined by the peaks of the signal waveform. Yet, it is the *average* voice power, which is some 10 dB to 20 dB below the peak amplitude, that determines the apparent loudness sensed by the ear—the ear does not hear peaks. Further, the normal dynamic range of human sounds varies from soft to strong, and often, the softer sounds are lost when they fall below the threshold of hearing.

Depending on the microphone used and the user's normal voice level, this microphone compressor will provide from 20 dB to 30 dB of signal compression (including peaks), ensuring a very high *talk power* for public-address systems, conference recordings, and amateur and Citizens-band transmitters. If you are skilled at making miniature printed-circuit boards, a small assembly, less batteries, can be built which will fit directly inside the larger variety of "stick" or "pencil" microphone. The unit can also be assembled on a perfboard and installed in a small aluminum enclosure. Each 12-volt power source can be two series-connected 6-volt batteries, such as the Eveready 2713 or the Burgess Z4. The compressor output should connect to a load impedance of 50,000 ohms or higher.

Any type of 741 integrated circuit can be used, though the half-minidip type "V" package is the easiest to wire on a perfboard.

A dynamic microphone of 50-, to 50,000-ohm impedance is suggested. If the unit is assembled in a metal enclosure, connect the microphone through a phone- or phono-type jack and plug.

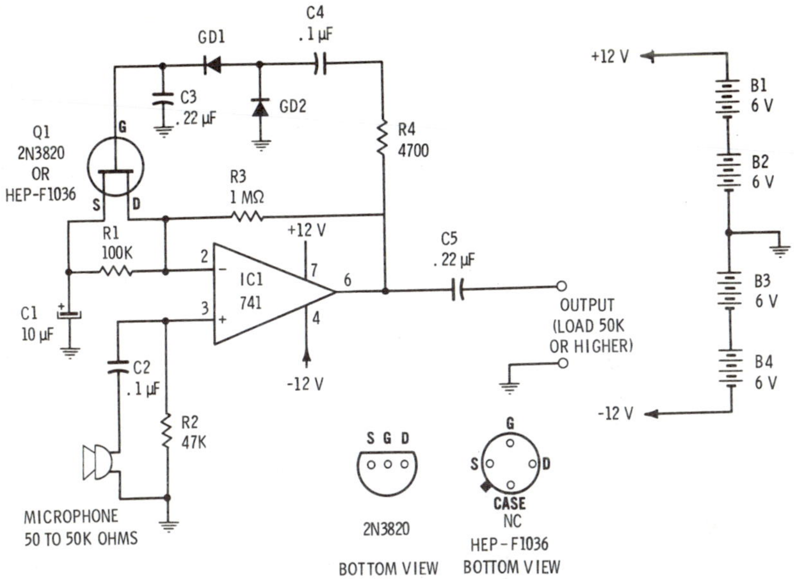

Microphone compressor.

PARTS LIST

Item	Description
R1	100,000-ohm, ½-watt resistor
R2	47,000-ohm, ½-watt resistor
R3	1-megohm, ½-watt resistor
R4	4700-ohm, ½-watt resistor
C1	10-μF, 6-V dc electrolytic capacitor
C2, C4	0.1-μF Mylar capacitor
C3, C5	0.22-μF Mylar capacitor
IC1	Type 741 operational amplifier (see text)
Q1	Type 2N3820, or Motorola HEP-F1036 FET
GD1, GD2	Germanium diode, type 1N60 or equivalent
Microphone	(See text)
B1, B2, B3, B4	6-volt battery (see text)

15. Magnetic Phono Stereo Preamplifier

With this easy-to-build stereo magnetic phono preamplifier, you can use a magnetic pickup to upgrade the sound of any public-address system presently using a "crystal/ceramic" record player. You can also upgrade small stereo sets, and even dub directly from your hi-fi turntable to a tape recorder. The specified IC contains two independent amplifiers with common power terminals. The terminals indicated in a circle are for the second amplifier, and the circuit shown must be built in duplicate (for both amplifier sections) for stereo operation. A simple bipolar power supply, as shown earlier in this book, should be used.

The specified parts values provide standard magnetic phono equalization. In a pinch, this preamplifier can be used for a stereo tape head since the equalization, although not perfect, will be adequate for all but the most precise applications.

The output is at line level—better than 1 volt. If the equipment following the preamplifier has an input volume control, resistor R6 can be replaced with a fixed 10K, ½-watt resistor.

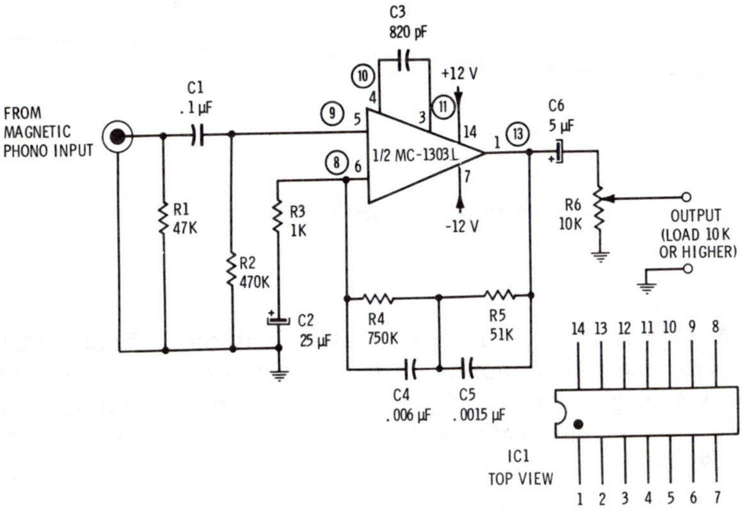

Magnetic phono stereo preamplifier.

PARTS LIST

Item	Description
R1	47,000-ohm, ½-watt resistor
R2	470,000-ohm, ½-watt resistor
R3	1000-ohm, ½-watt resistor
R4	750,000-ohm, ½-watt resistor
R5	51,000-ohm, ½-watt resistor
R6	10,000-ohm, audio-taper potentiometer (see text)
C1	0.1-μF Mylar capacitor
C2	25-μF, 6-V dc, or higher, electrolytic capacitor
C3	820-pF disc capacitor
C4	0.006-μF disc capacitor (0.005 + 0.001)
C5	0.0015-μF disc capacitor
C6	5-μF, 15-V dc, or higher, electrolytic capacitor
IC1	Motorola MC 1303L

16. Back-of-Speaker Amplifier

Assemble this amplifier on a small perfboard about 1 inch square, cement the perfboard to the back of a small speaker, and you've got a monitor amplifier that can be built into portable recorders or into "amateur-type" microphone mixers and consoles. Volume control R1 can be a small trimmer potentiometer located right on the perfboard; if you want full-time control, R1 can be a standard-size potentiometer mounted on the front of the recorder or mixer.

Speaker SPK1 can be any size, 2 inches or larger, such as those sold as "transistor-radio replacements." Don't worry about wires sticking through the perfboard and being shorted by the back of the speaker—the magnet. Just use a large blob of RTV adhesive, such as *Silastic* or bathtub caulking.

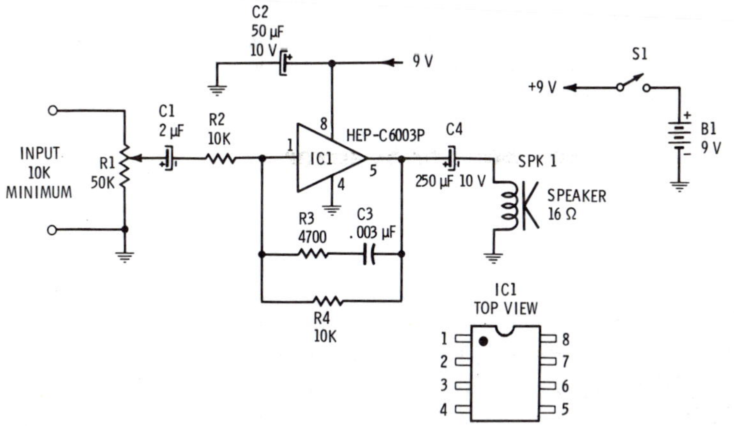

Back-of-speaker amplifier.

PARTS LIST

Item	Description
R1	50,000-ohm trimmer or standard potentiometer
R2, R4	10,000-ohm, ½-watt resistor
R3	4700-ohm, ½-watt resistor
C1	2-μF, 6-V dc, or higher, electrolytic capacitor
C2	50-μF, 10-V dc, or higher, electrolytic capacitor
C3	0.003-μF, 10-V dc, or higher, capacitor
C4	250-μF, 10-V dc, or higher, capacitor
SPK1	16-ohm speaker
IC1	Motorola HEP-C6003P
B1	9-volt battery, transistor-radio type
S1	Switch, spst

17. Low-Noise Microphone Preamplifier

By using a specially designed low-noise dual preamplifier, you can assemble a microphone mixer that has full professional characteristics. In the circuit shown, the power-supply connections (pins 4 and 11) are common to both preamplifiers. The terminal numbers indicated in the rectangles are for the second preamplifier section.

The frequency response is essentially flat from 20 to 20,000 Hz at a distortion no higher than 0.2% thd at any frequency. If the output load is 10K ohms or higher, the worst-case noise figure is 1.6 dB. The circuit input impedance is 100,000 ohms, and any microphone with a 50- to 50,000-ohm impedance can be connected directly to the amplifier. The stage gain is nominally 40 dB.

Note that the circuit requires a single-ended power supply capable of providing a maximum of 16 mA of current (8 mA per section). Also note there is no connection to the inverting input (pins 1 or 13). For maximum stability, capacitors C2 and C3 should be the tantalum type, though it is possible to reduce costs by using a good-grade ordinary electrolytic for C3. Capacitor C4 must be used, preferably installed directly at the IC terminal. If two or more dual preamplifiers are used in a multimicrophone mixer, it might be necessary to increase the value of C4 to 250 μF, or even 500 μF, depending on the number of inputs. Low-frequency "motorboating" is an indication that the value of C4 should be increased by at least a factor of 2.5. If there is evidence of high-frequency oscillation, bypass C4 with a 0.1-μF Mylar or ceramic capacitor directly at the IC terminal.

Although the circuit is intended for operation into a mixing bus, you can use it as a straight preamplifier by connecting the outputs into the "line" inputs of a reel-to-reel, cartridge, or cassette recorder.

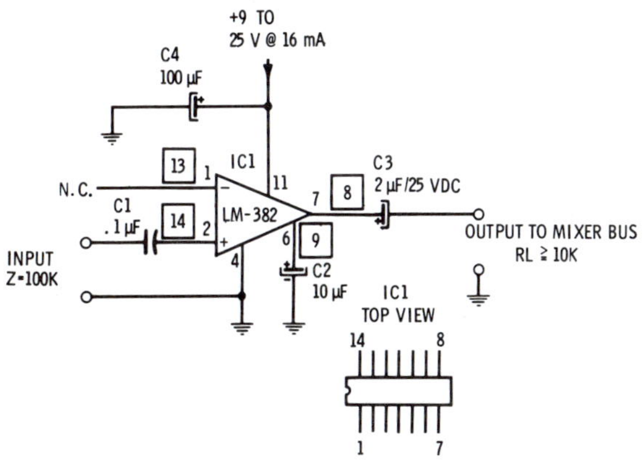

Dual low-noise microphone preamplifier.

PARTS LIST

Item	Description
C1	0.1-μF Mylar capacitor
C2	10-μF, 3-V dc tantalum capacitor
C3	2-μF, 25-V dc tantalum or electrolytic capacitor (see text)
C4	100-μF, 25-V dc electrolytic capacitor (see text)
IC1	National LM382

[illegible]

Item	Description
C1	[illegible] Mylar capacitor
C2	[illegible] aluminum electrolytic [illegible]
C3	[illegible]
[illegible]	[illegible]

Filters and Oscillators

18. 60-Hz Hum Buster

It happens quite often. You borrow a tape for dubbing or you make a live recording, and on playback you find that somehow there is almost as much hum as there is program signal. This also happens in recording studios when tapes from outside sources are used or when a remote recording is done too close to power-line circuits. The studios get rid of the hum by passing the signal through a notch filter, a device that removes a very narrow band of frequencies while passing all others without attenuation. At the 60-Hz hum frequency, very little of the immediate adjacent frequencies are removed—only the hum.

This notch filter can do almost as good a job as many studio filters and it can be assembled by using almost any surplus operational amplifier. For maximum flat response, a 531 is suggested, although a 741 works almost as well in terms of frequency response. If you use anything other than a compensated IC such as the 741, make certain you provide the required compensation components to prevent high-frequency oscillation. (See the manufacturer's specifications.)

Resistors R1 and R2 should have 1% tolerance. Resistor R3 can have the nearest standard 5%-tolerance value. Capacitors C1, C2, and C3 should have 1% tolerance, although you can use 5% units and trim them to a matched value by using a capacity bridge or meter.

Potentiometer R4 is the "Q" control; normally it is set so that the wiper contact is grounded. Resistor R4 should be adjusted to provide the maximum hum filtering coincident with the least attenuation of other bass frequencies. It can also be used to filter hum that is not precisely 60 Hz by broadening the bandwidth.

The unit should be assembled in a metal enclosure. A standard transistor-radio-type battery supply can be used and is suggested. If an ac power-line supply is used, make certain its filtering is suitable for use with a microphone preamplifier; there's no sense adding hum after it has been filtered out.

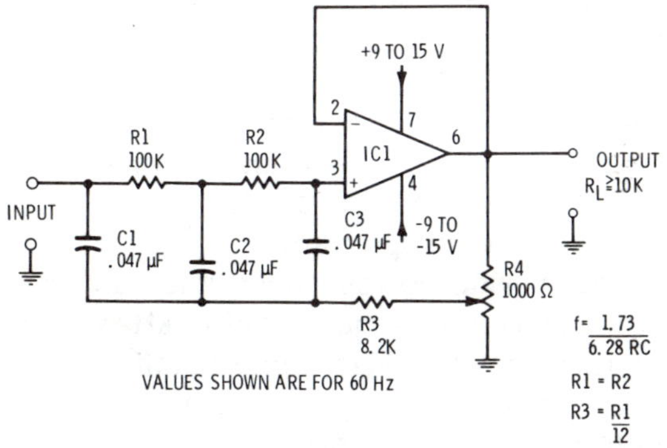

A 60-Hz hum buster.

PARTS LIST

Item	Description
R1, R2	100,000-ohm, ½-watt, 1% resistor
R3	8200-ohm, ½-watt, 5% resistor (see text)
C1, C2, C3	0.047-µF capacitor (see text)
R4	1000-ohm potentiometer
IC1	See text

19. Low-Pass Filter

The easiest way to separate a desired signal from noise is to use some sort of filtering device that will pass the desired signal—or most of it—while attenuating the noise. For hobbyists, this type of circuit is the "low-pass filter;" it is usually termed a "high filter," or "hi-filter," when used in high-fidelity amplifiers. Low-pass filters are also common to differential oscilloscope preamplifiers, where low-frequency phenomenon is often obscured by high-frequency noise. Removing the noise generally produces a clear scope trace.

In the circuit shown, the resistance and capacitance values indicated produce a cutoff frequency of 10 kHz which attenuates 12 dB per octave above 10 kHz. Other cutoff frequencies, such as 7000 Hz for removing "scratch" from 78-rpm records, can be obtained from the formula shown.

Resistors R1 and R2 are equal and can be any convenient value, such as 47,000 ohms or 100,000 ohms. The idea is to use common capacitor values and calculate for the proper resistance, which can be constructed from two or three standard resistance values. Note that capacitor C2 is always twice the value of C1.

Also note there is no *inverting input* on the specified IC; the connection to the output is internal. If you substitute another type of operational amplifier such as a 741 or 709, be certain you make the connection indicated by the dotted line.

The power supply can be constructed from ordinary transistor-radio-type batteries, or a line-powered supply can be used. If you use a line-powered supply, make certain it is well filtered.

The LM310 IC is available in several configurations. The round, or metal, package has the terminal connections indicated. Other packages do not necessarily conform to standard terminal arrangements, and you should specifically check the connections for any other type of package.

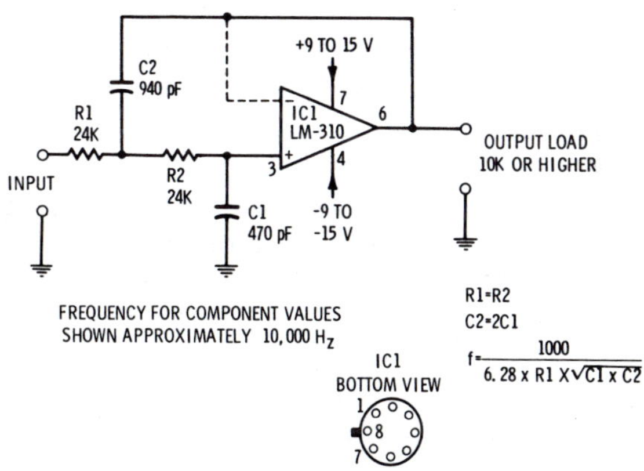

Low-pass filter.

PARTS LIST

Item	Description
R1, R2	See schematic and text
C1, C2	5% silver mica capacitor (see schematic and text)
IC1	National LM310

20. High-Pass Filter—12 dB/octave

A high-pass filter attenuates those frequencies falling *below* the cutoff frequency. In high-fidelity amplifiers a high-pass filter is generally termed a "low filter," or "lo filter." In the circuit shown, the indicated resistance and capacitance values provide a cutoff frequency of approximately 100 Hz, suitable for a turntable rumble filter. Other cutoff frequencies can be calculated by using the formula shown.

Capacitors C1 and C2 are equal and can be any convenient value. Resistor R2 is twice the value of R1. In this circuit it is easier to assign standard values to the resistors and calculate the capacitor value, which can be obtained by parallel-connecting two or three standard capacitor values. However, the 5% silver mica capacitors, which should be used, are somewhat expensive; so see if you can use standard capacitor values, such as 0.022 μF, and *trim* the resistors.

Note there is no *inverting input* on the specified IC; there is an internal connection to the output. If you substitute another type of operational amplifier, such as the 741 or 709, be certain you make the connection indicated by the dotted line.

The power supply can be ordinary transistor-radio-type batteries or a line-powered supply. If the supply is line-powered, be certain it has filtering suitable for use with a microphone preamplifier. There is no sense in adding hum after you have filtered it out from the signal source.

The LM310 IC is available in several configurations. The round, or metal, package has the terminal connections indicated. Other packages do not necessarily conform to standard terminal arrangements, and you should specifically check the connections for any other type of package.

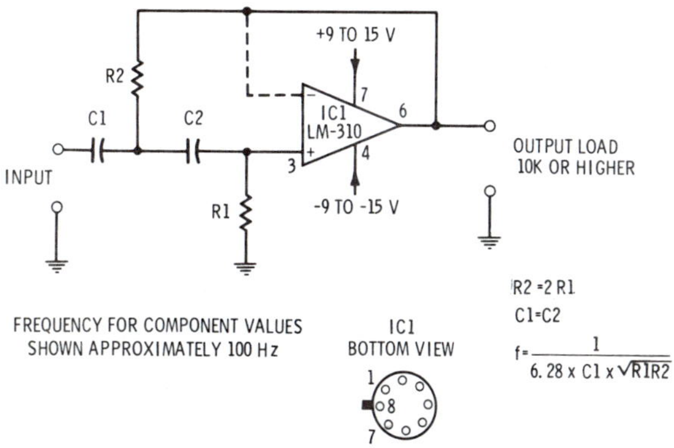

High-pass filter.

PARTS LIST

Item	Description
R1, R2	See schematic and text
C1, C2	5% silver mica capacitor (see schematic and text)
IC1	National LM310

21. Cue-Tone Oscillator

Today, we used tone control for remote operation of telephone answering devices, for Citizens band and amateur radio signaling and transmitter control, and even for model airplane and boat control. This cue-tone oscillator, which can be used for any type of tone control, can be assembled in a pocket-sized plastic box and powered by ordinary transistor-radio batteries. Any frequency up to approximately 10 kHz can be attained by using the formula shown. When possible, the value for "R" should be equal to or less than 47,000 ohms for maximum frequency stability. Mica or tantalum capacitors are suggested for "C." Depending on the frequency, the maximum output voltage will be nominally 5 V rms.

The rough or coarse output voltage is set with potentiometer R1, which can be a small trimmer type. Make certain R1 is set so that the oscillator starts immediately when power is applied. The fine output voltage is determined by potentiometer R2.

Operational amplifier IC1 can be any model of the type 741.

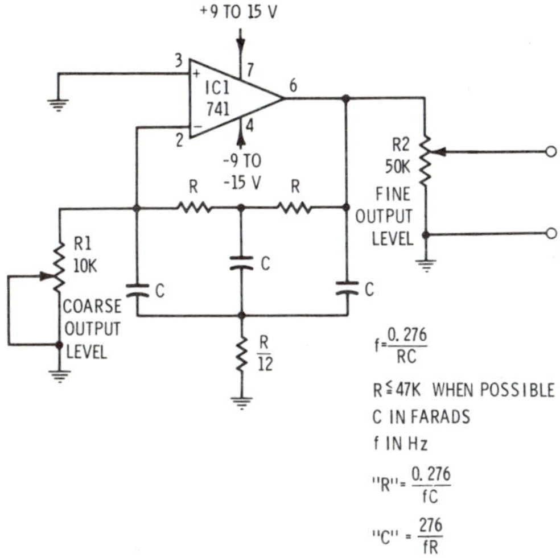

Cue-tone oscillator.

PARTS LIST

Item	Description
R1	10,000-ohm potentiometer (trimmer type)
"R"	See text
R2	50,000-ohm potentiometer
"C"	See text (10-V dc or higher)
IC1	Operational amplifier type 741

22. Cue-Tone Amplifier

This cue-tone amplifier is the "receiver" for the cue-tone oscillator. Several amplifiers can be connected across the signal source to receive two or more simultaneous cue tones. Note that the circuit is very similar to the cue-tone oscillator, although a maximum value of 27,000 ohms is suggested for "R." Again, transistor-radio-type batteries can be used for the power supply, while IC1 can be any model of the type 741 operational amplifier.

Potentiometer R2, the "Q" control, sets the bandpass of the amplifier. If the component values are not precise, a low "Q" would be used so that the incoming frequency could fall within a moderately wide frequency range. If component values are precise, a high "Q" can be used for a narrow bandpass so that two closely spaced cue-tone frequencies can be separated.

The input signal level should be approximately 1-V rms or lower.

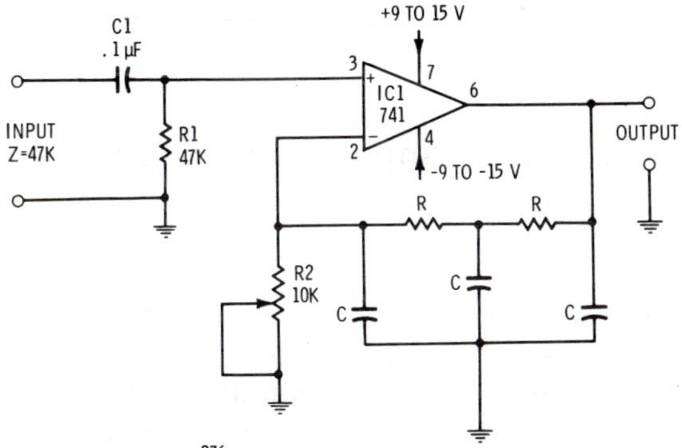

$$f = \frac{276}{RC}$$

R $\leq$ 27K WHEN POSSIBLE

C IN FARADS

f IN Hz

$$\text{"R"} = \frac{0.276}{fC}$$

$$\text{"C"} = \frac{0.276}{fR}$$

Cue-tone amplifier.

PARTS LIST

Item	Description
R1	47,000-ohm resistor
"R"	See text
R2	10,000-ohm potentiometer
C1	0.1-μF Mylar capacitor
"C"	10-V dc, or higher, mica or tantalum capacitor
IC1	Operational amplifier type 741

23. High-Output Audio Oscillator

The next time you need a low-distortion signal source to check out a tape recorder, amplifier, or home-brewed audio project you can throw this high-output oscillator together in a few minutes from junk-box parts. Depending on the setting of level control R6, the circuit delivers approximately a 1000-Hz signal at a few millivolts to better than 5 volts rms.

Integrated circuit IC1 can be just about any operational amplifier, although the type 741 is the easiest to use. If you use a high-frequency IC such as the NE531, connect the compensation capacitor, C_X, as indicated in the schematic. Note carefully that the IC positive (noninverting) input is grounded.

The circuit can be powered by two 9-volt transistor-radio-type batteries as shown. If the batteries are old in the sense they've been on the shelf for several months, it might be necessary to bypass each battery with a 50-μF or 100-μF capacitor. The need for the capacitors is indicated if the oscillator fails to start or if the output signal is weak, intermittent, or highly distorted.

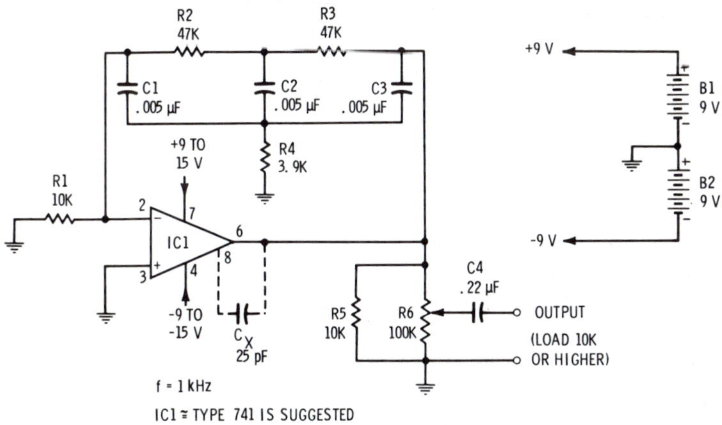

High-output audio oscillator.

PARTS LIST

Item	Description
R1, R5	10,000-ohm, ½-watt resistor
R2, R3	47,000-ohm, ½-watt, 5% resistor
R4	3900-ohm, ½-watt, 5% resistor
R6	100,000-ohm potentiometer
C1, C2, C3	0.005-μF capacitor (5% suggested)
C4	0.22-μF, 25-V dc capacitor (Mylar suggested)
Cx	25-pF, ceramic disc capacitor (see text)
IC1	Operational amplifier (see text)
B1, B2	9-volt transistor-radio-type battery

Hobby Circuits

24. SCA Adapter

In many communities, one or more fm stations provide a background music service for stores and restaurants in addition to the regular (main channel) programs. While a standard fm radio receives only the normal, or main channel, signal, a receiver equipped with a special decoder called an *SCA adapter* can receive the background music, which is generally free of commercial interruptions.

The circuit shown can be assembled on a section of perfboard using push-in terminals for tie points. Parts placement is not critical; however, connect the ground leads of C6, C7, and C8 to the same point and then connect this tie point to ground. If the audio output is of extremely low level and very *muddy,* lift the ground leads from these three capacitors and connect them directly to the positive voltage of the power supply, which appears at IC pin 10.

Potentiometer R7 is a trimmer type; there is no need for a standard potentiometer because it generally does not require a precise adjustment or frequent resetting.

The SCA adapter connects to the detector output of a mono or a stereo receiver before the de-emphasis network or before the SCA trap which is used ahead of some stereo multiplex demodulators. If you connect the adapter after the de-emphasis network, you will find the background music subcarrier has been filtered and all you will get out of the adapter is noise. Many modern fm tuners and receivers have an output jack labeled "fm detector" or "fm 4-channel;" these jacks are simply the fm detector output before the de-emphasis filter and provide the signal needed for the SCA adapter.

The SCA adapter requires an input signal of at least 100 mV from the fm detector. If your fm radio or receiver cannot deliver this signal level, connect the booster amplifier shown, between the fm detector and the SCA adapter. The booster can be assembled on a small section of perfboard, and there is no need to build it until you are certain that it is needed. Normally, the booster output can connect to IC1 pin 2 terminal, and C1, C2, and R1 can be eliminated. If you have assembled the SCA adapter, however, eliminate C106 (the 470-pF booster output capacitor) and connect the booster output to C1.

To locate an SCA broadcast, set the R7 wiper to the mid position, connect a high-gain audio amplifier to the adapter output, and tune across the fm band. When you hear modulation suddenly appear in the monitoring speaker, carefully tune the station and then adjust R7 for best sound. Keep in mind that the frequency response of background music programming is sharply limited on the high end (above 5000 Hz), so don't expect hi-fi. If R7 cannot clean up the signal, check to be certain you are not hearing the main stereo channel; there is often "stereo spillover" which at first appears to be distorted SCA programming. If you are certain that what you are hearing is background music, but it is weak, noisy, and/or distorted, you most likely need to use the booster amplifier.

PARTS LIST

Item	Description
	Adapter
R1, R3, R4, R5	4700-ohm, ½-watt resistor
R2	10,000-ohm, ½-watt resistor
R6	1800-ohm, ½-watt resistor
R7	5000-ohm trimmer potentiometer
R8, R9, R10	1000-ohm, ½-watt resistor
C1, C2	510-pF ceramic disc capacitor
C3	1000-pF (0.001-μF) ceramic disc capacitor
C4, C5	0.001-μF Mylar capacitor
C6, C8	0.02-μF Mylar capacitor
C7	0.047-μF (or 0.05) Mylar capacitor
IC1	NE565 phase-locked loop, "A" package suggested
	Booster
R101	4700-ohm, ½-watt resistor
R102, R105	47,000-ohm, ½-watt resistor
R103	10,000-ohm, ½-watt resistor
R104	470-ohm, ½-watt resistor
C101	470-pF, 5% capacitor
C102	47-pF, 5% capacitor
C103	0.01-μF Mylar capacitor
C104	0.1-μF Mylar capacitor
C105	10-pF ceramic disc capacitor
C106	470-pF ceramic disc capacitor (see text)
IC2	SE or NE531 operational amplifier

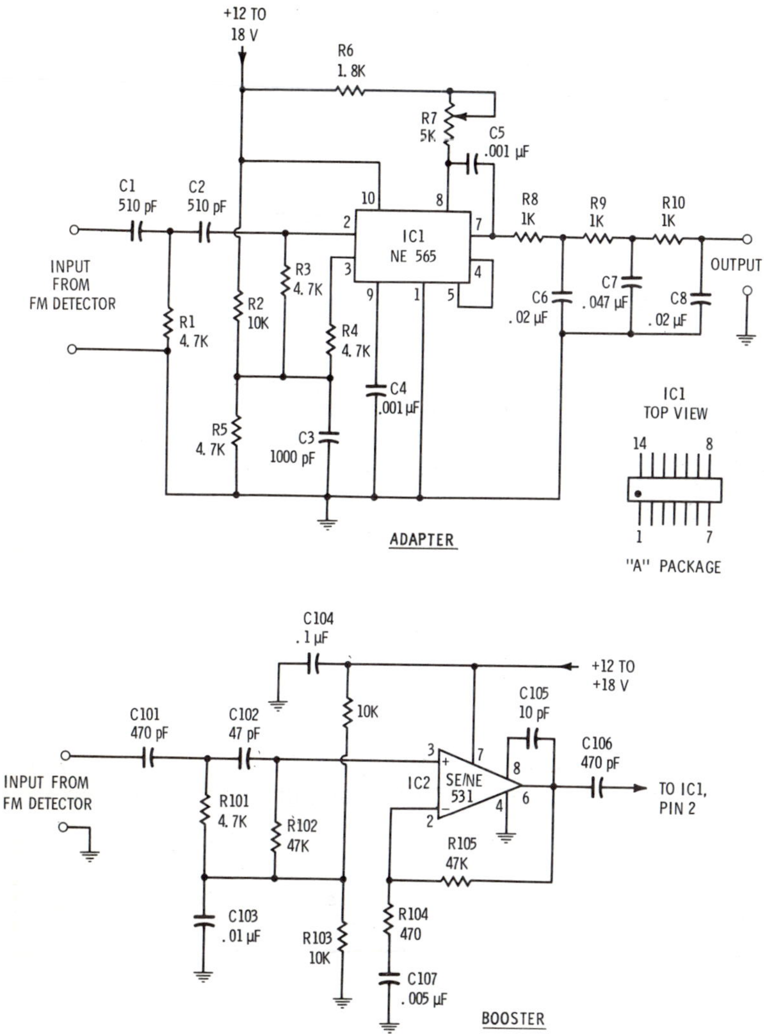
+12 TO 18 V
R6 1.8K
R7 5K
C5 .001 µF
C1 510 pF
C2 510 pF
10
8
2
IC1 NE 565
7
3
4
9
1
5
R8 1K
R9 1K
R10 1K
INPUT FROM FM DETECTOR
R3 4.7K
R2 10K
R1 4.7K
R4 4.7K
C7 .047 µF
C6 .02 µF
C8 .02 µF
OUTPUT
C4 .001 µF
R5 4.7K
C3 1000 pF
IC1 TOP VIEW
14
8
1
7
"A" PACKAGE
ADAPTER
C104 .1 µF
+12 TO +18 V
10K
C105 10 pF
C101 470 pF
C102 47 pF
C106 470 pF
3
7
8
IC2
SE/NE 531
6
4
2
TO IC1, PIN 2
INPUT FROM FM DETECTOR
R101 4.7K
R102 47K
R105 47K
R104 470
C103 .01 µF
R103 10K
C107 .005 µF
BOOSTER

SCA adapter.

25. Enlarger Timer

This electronic timer will give precise exposure control for photographic enlargers and printers. The working range is 1 to 60 seconds, the exact time interval established by potentiometer R2. The timer closes relay RY1 the instant that normally open push switch PB1 is closed. The B and C contacts of relay RY1 then close. If the enlarger lamp is connected to contacts B and C, the lamp will remain on for as long as the timer is programmed. When the timer turns off, the B and C contacts open, extinguishing the lamp. A safelight can be powered through contacts A and B; the safelight will automatically turn off when the enlarger is turned on.

Relay RY1 can be any 6-volt dc type with a coil resistance of 100 ohms or higher.

The simple power supply shown, using a standard 12.6-volt filament transformer rated 500 mA or higher, can be used to power the timer.

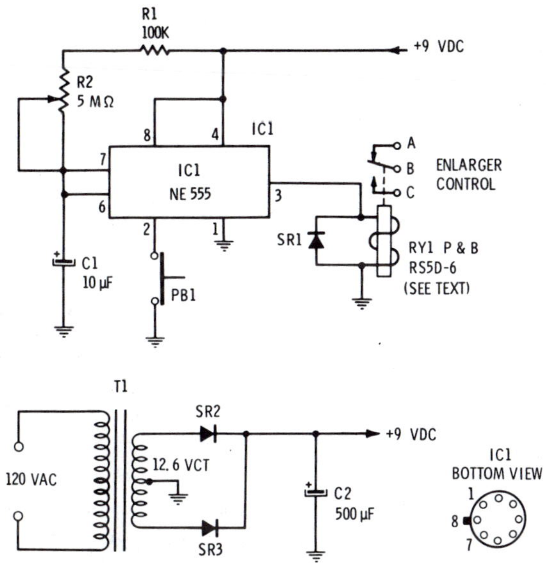

Enlarger timer.

PARTS LIST

Item	Description
R1	100K, ½-watt resistor
R2	5-megohm linear-taper potentiometer
C1	10-µF, 15-V dc tantalum capacitor
C2	500-µF, 15-V dc electrolytic capacitor
IC1	NE 555 timer IC
RY1	Potter and Brumfield RS5D-6 relay or equivalent (see text)
SR1	Silicon diode, 100 PIV, 0.5 A
SR2, SR3	Silicon diode, 25 PIV, 0.5 A or higher
T1	Filament transformer, 12.6-VCT, 0.5 A (see text)
PB1	Normally open spring-return push-button switch

26. Timed Auto Burglar Alarm

Many communities now restrict the length of time a home and auto burglar alarm can continuously ring, and the list of towns and cities passing ordinances to restrict noise pollution grows weekly.

This auto burglar alarm is timed so that after a fixed interval—long enough to frighten away a burglar—the alarm bell is automatically silenced until the alarm is reset by the user. Though intended primarily to protect motor vehicles, it is easily adapted to home use by simply using a 12-volt lantern battery (or two 6-volt series-connected types) as the power source.

A fender-mounted keyswitch arms the circuit and turns it off, providing a reset in the off-on sequence. The protective switches are the OC (open-circuit) type sold in auto supply stores for auto burglar alarms. These spring-loaded switches are normally closed but are held open by the force of the closed door. When a door is opened, a switch closes to trigger the alarm. Relay RY1 latches so that the alarm continues to ring even if the protective switch is restored to its open condition.

The length of time the alarm will continue to sound after being triggered is approximately equal to R1 × C1 (farads). The combination shown—100K and 500 μF—will provide a continuous alarm of approximately 1 minute. Doubling the resistance and capacitance values will provide a continuous alarm of approximately 2 minutes.

To avoid unreliable operation caused by excess ambient heat, you should install the alarm circuit in the passenger compartment, rather than under the hood where it would be subjected to engine heat.

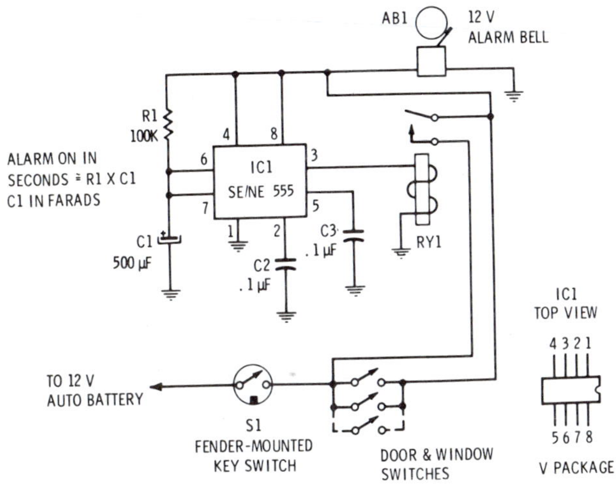

Timed auto burglar alarm.

PARTS LIST

Item	Description
R1	100,000-ohm, ½-watt resistor (see text)
C1	500-μF, 25-V dc electrolytic capacitor (see text)
C2, C3	0.1-μF, 25-V dc, or higher, capacitor
RY1	Relay, 12-V dc, Potter & Brumfield RS5D-12 or equivalent
IC1	Timer, type NE or SE 555
S1	Keyswitch
AB1	12-V dc alarm bell or siren
Miscellaneous	Protective switches for doors (see text)

27. Audio Signal Tracer

If you use subminiature parts, you can probably assemble the entire high-gain audio signal tracer, except the speaker, in an old pocket flashlight. And, if you replace the speaker with a low-impedance (8 to 100 ohms) earphone, you can keep the whole thing in your shirt pocket. If you prefer using standard components (to keep costs at rock bottom), you can assemble everything in a small aluminum cabinet approximately 3 × 4 × 5 inches. Use the largest speaker that will fit into the enclosure because the larger the speaker, the better the low-frequency response.

The overall gain is high enough to permit monitoring directly from a microphone or from the output of a magnetic phono pickup. There are no critical assembly procedures. Since the circuit is not sensitive to high frequencies above the audio range, the most compact arrangement is through a small printed-circuit board. The power supply can be any small 9-volt transistor-radio battery.

Maximum power output is attained from a 16- or 22-ohm speaker, but you can use a 40-ohm intercom speaker with not too much in the way of reduced power output. Transistor Q1 can be a silicon pnp, such as the 2N5355 or equivalent. The exact lead configuration of Q1 will depend on the particular transistor used.

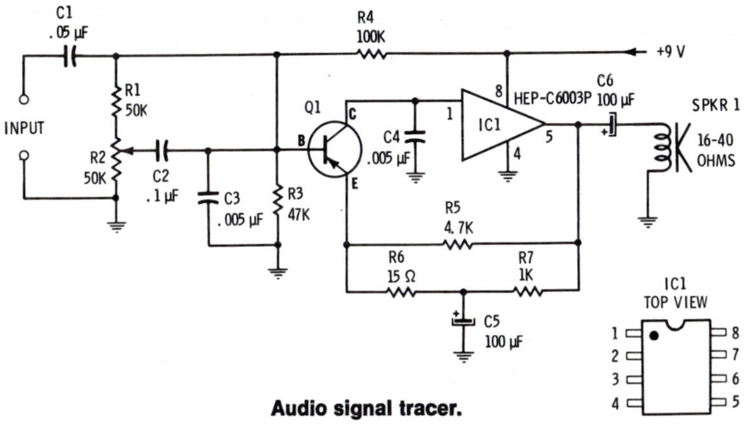

Audio signal tracer.

PARTS LIST

Item	Description
R1	50,000-ohm, ½-watt resistor
R2	50,000-ohm audio-taper potentiometer
R3	47,000-ohm, ½-watt resistor
R4	100,000-ohm, ½-watt resistor
R5	4700-ohm, ½-watt resistor
R6	15-ohm, ½-watt resistor
R7	1000-ohm, ½-watt resistor
C1	0.05-μF, 500-V dc ceramic disc capacitor
C2	0.1-μF, 100-V dc Mylar capacitor
C3, C4	0.005-μF ceramic disc capacitor
C5, C6	100-μF, 10-V dc electrolytic capacitor
SPKR1	16- to 40-ohm speaker (see text)
IC1	Motorola HEP-C6003P
Q1	Transistor, 2N5355 or equivalent

28. Stereo Balancer

Optimum sound quality is obtained from your stereo system when the volume and the tone between the channels are in precise electrical balance. This differential meter will indicate the balanced condition and can be left permanently connected because it has no effect on final sound quality.

If you have a VU meter handy, use the circuit arrangement shown. If you do not have a VU meter (they are expensive), you can substitute a 0-100–dc microammeter with a silicon diode bridge, as shown in the alternate schematic.

Meter M1 will indicate only when there is a voltage difference at the input terminals, which are connected across the amplifier speaker terminals.

To adjust the amplifier for balance, set the amplifier stereo-mono switch to *mono* so that both speakers receive exactly the same program signal, or use a mono signal source such as a mono fm station. Adjust the amplifier balance control for a zero meter reading, or as close to zero as you can get. Then adjust the amplifier tone controls for a null or zero meter reading. These adjustments are most easily done if the signal source is a sustained tone, such as the kind available on "consumer test records."

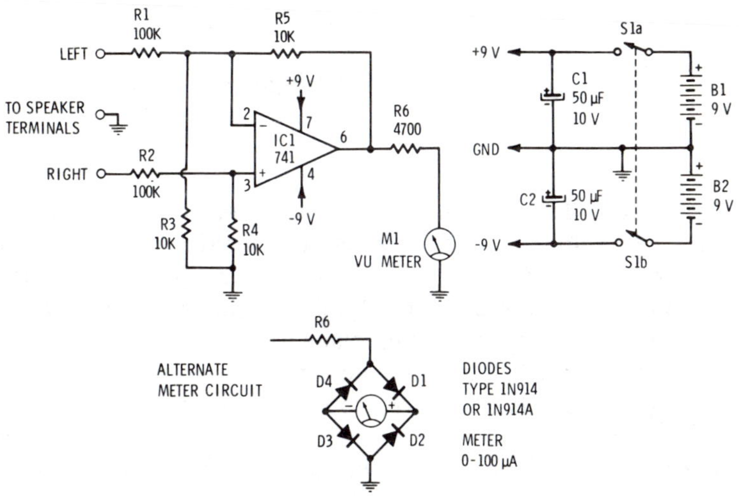

Stereo balancer.

PARTS LIST

Item	Description
R1, R2	100,000-ohm, ½-watt resistor
R3, R4, R5	10,000-ohm, ½-watt resistor
R6	4700-ohm, ½-watt resistor
C1, C2	50-μF, 10-V dc, or higher, electrolytic capacitor
IC1	Type 741 operational amplifier
M1	Miniature VU meter (see text)
B1, B2	9-volt transistor-radio battery, Eveready 916 or equivalent
D1, D2, D3, D4	Silicon diode, type 1N914 or 1N914A (see text)
M1 Alternate	Meter, 0-100–dc microammeter (see text)
S1	Switch, dpst

29. Light Meter

Although this circuit appears unnecessarily complex for a photo light meter, its supersensitivity makes it ideal for use as an "enlarging meter." It can even read light values through Wratten filters for color printing. The power supply can be a simple bipolar type shown elsewhere in this book. Nothing about the circuit or assembly is critical.

A typical use is to ensure proper printing exposure. After you have made a good print by using a 10-second (or 20-second) exposure, leave the enlarger controls alone and place a diffuser under the lens. Position photocell PR1 on the easel directly under the lens and adjust R3 for a center-scale meter reading. Note the setting of R3. After you compose, using a new negative, or after you change the enlarging ratio, position the diffuser under the lens, place PR1 on the easel, and adjust the lens opening (diaphragm) for a center-scale reading. The exposure will be set for your standard printing (exposure) time. Do not worry if the meter *pins* when room lights are turned on, there will be no damage.

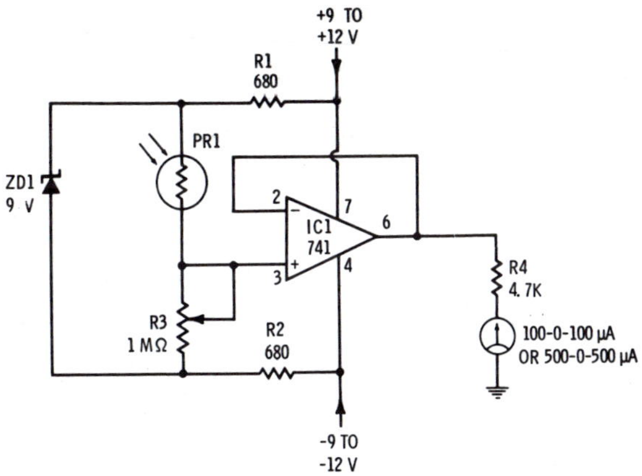

Light meter.

PARTS LIST

Item	Description
R1, R2	680-ohm, ½-watt resistor
R3	1-megohm potentiometer
R4	4700-ohm, ½-watt resistor
IC1	Type 741 operational amplifier
ZD1	9-volt, 1-watt zener diode
PR1	Photoresistor, Clairex CL5M5
M1	100-0-100–μA (preferred) or 500-0-500–μA meter
(Note: R4 can be adjusted to provide any desired standard setting for R3.)	

30. Portable Intercom

This battery-powered intercom takes only one integrated circuit, yet it packs as much talk-power as many of the commercial intercoms sold in department stores.

The power supply can be either a single 6-volt lantern battery, two series connected 6-volt batteries, a 12-volt lantern battery, or an automobile battery. Regardless of the power supply used, an external filter capacitor is not required because it is built into the circuit as C1. A 12-volt supply is suggested where maximum talk-power is required because the output is approximately 4 times that attained with a 6-volt supply.

If a spring-return switch is used for S1, the master speaker (SPKR 1) will continuously monitor the remote speaker (as a "baby sitter"). When S1 is depressed, the master speaker can communicate with the remote speaker. Releasing S1 will automatically restore monitoring of the remote.

A heat sink is not required if the speakers are 16 ohm. If you substitute 8-ohm or 4-ohm speakers, a "piggyback" finned heat sink must be attached to IC1.

Transformer T1 can be any kind of "transistor" transformer with 16-ohm and 1000-ohm windings. Since the turns ratio is approximately 8:1, even an 8- to 500-ohm transformer can be used. Use whatever is available at the lowest possible cost from any source. A "surplus" transformer is, of course, the best value.

Components labeled CX and RX are used only if the circuit breaks into self-oscillation without them. Try the circuit before installing CX and RX. If you can accept a sharply reduced bass response, C2 can be reduced to 100 μF.

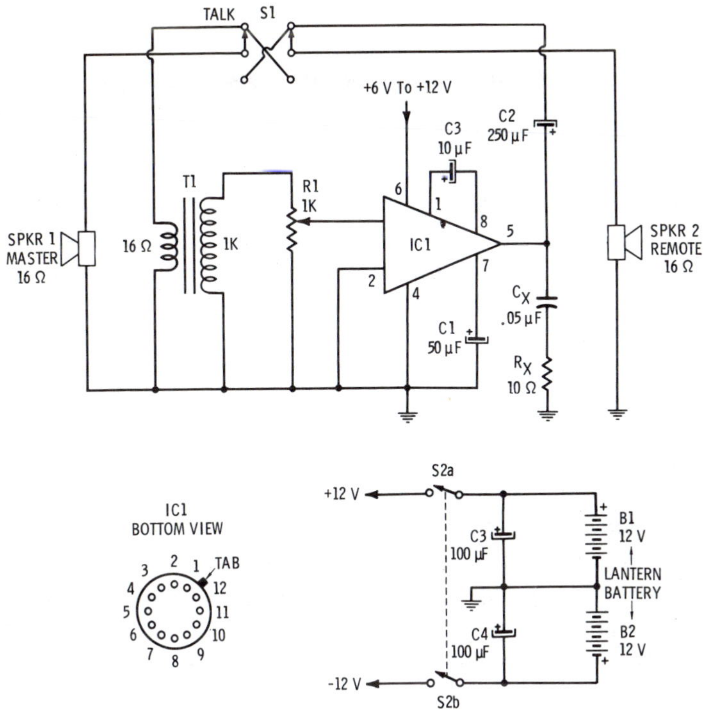

Portable intercom.

PARTS LIST

Item	Description
C1	50-μF, 15-V dc, or higher, capacitor
C2	250-μF, 15-V dc, or higher, capacitor (see text)
C3	10-μF, 15-V dc, or higher, capacitor
CX	0.05-μF, ceramic disc (see text)
R1	1000-ohm potentiometer
RX	10-ohm, ½ watt resistor (see text)
T1	16-ohm to 1000-ohm transformer (see text)
SPKR 1	16-ohm speaker
SPKR 2	16-ohm speaker
IC1	LM386, or Radio Shack 276-1731
S1	Switch, dpdt

31. Regulated Op-Amp Power Supply

Although the unregulated bipolar power supply generally suffices for most hobbyist applications of operational amplifier projects, there are occasions when you might require a regulated power supply. Perhaps you have built a precision oscillator and the local line voltage wanders between 105 and 135 volts, or maybe the current requirements of your project vary over a relatively wide range, causing an unregulated dc supply voltage output to decrease as the current increases. Whatever this regulated supply will keep the dc voltage supply rock-steady over a wide range of line-voltage variations at output currents up to 1 ampere.

Note carefully that IC1 and IC2 have different connections, even though they appear to be the same. Check their identification numbers very carefully before installation. The negative regulator, IC2, uses pin number 2 for the input, in contrast to the commonly used pin number 1 for positive regulators. Both ICs must be secured to a heat sink or the chassis to attain the full 1-ampere capability. If you only require up to 100-mA output current, a 78L12 can be substituted for IC1 and a 79L12 can be substituted for IC2. These regulators are very small, do not require a heat sink, and have the same pin terminal connections. Transformer IC1 should have a secondary current rating at the very least equal to the maximum current rating of the regulators although T1 will run much cooler if the current rating is 150% of the maximum required by the project.

No special construction precautions are necessary other than keeping C3 and C4 as close to the ICs as is possible.

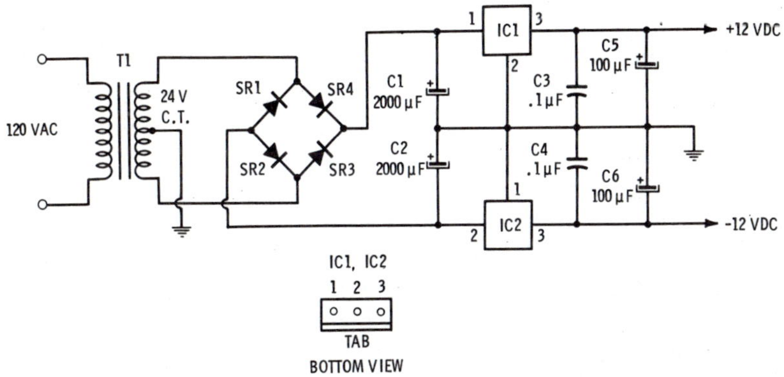

Regulated op-amp power supply

PARTS LIST

Item	Description
C1, C2	2000-μF, 25-V dc electrolytic capacitor
C3, C4	0.1-μF Mylar capacitor
C5, C6	100-μF, 15-V dc electrolytic capacitor
IC1	Positive voltage regulator, 7812 (see text)
IC2	Negative voltage regulator, 7912 (see text)
SR1 to SR4	Silicon rectifier, 50 PIV, 1 A.
T1	120/24 V C.T. transformer, rated at least 1A (see text)
Miscellaneous	Chassis or heat sink if needed (see text)

32. Hi-Level Mixer

Sometimes luck just strikes, and it is possible simply to tie the output connections of several tape recorders and line amplifiers together and feed the mixed product into yet another recorder, when producing a "master sound tape." Other times, the machines simply reject each other and the end product is howling, whistling, or just plain distortion. The easiest way to insure clean sound mixing from high-level sources, such as tape recorders, turntable preamplifiers, and line amps is to combine their outputs through a high-level mixer, or *summing amplifier* as it is more often known.

This project is intended for solid-state and other low-impedance recorders, amps, and preamps—meaning equipment with output impedances less than 600 to 1000 ohms—which includes most solid-state equipment.

There is nothing critical about construction other than placing C1 and C2 as close as possible to the terminals of IC1. The exact pinout will be determined by the type of IC package. Most of the common pinouts are shown throughout this book.

R1, R2, etc., represent the total number of active inputs. If you have five inputs, each input connects through a 100K resistor (R_n) to IC1 pin 2.

Resistor R4 is only a protection against accidental shorting IC1 output to ground. If R_L is greater than 10,000 ohms, as it generally will be if you use a "bridging" (high-impedance) recorder input, R4 can be replaced with 0.47-μF capacitor rated 15 V dc, or higher.

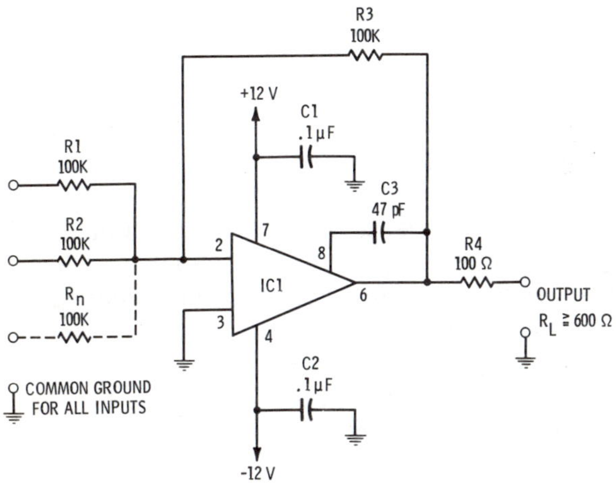

Hi-level mixer

PARTS LIST

Item	Description
R1, R2, R3	100K, ½-watt resistor, 1% low-noise type suggested
R_n	Same as R1 (see text)
C1, C2	0.1-μF Mylar capacitor
C3	47-pF ceramic disc capacitor
IC1	Signetics NE or SE 531 operational amplifier
Miscellaneous	R4 (100 ohms) if needed (see text)

33. Scope Calibrator

Most low-cost general-purpose oscilloscopes have uncalibrated horizontal sweep selectors, yet today an accurate time-base measurement is often required for proper adjustment of turntables, timing devices, and digital equipment.

Utilizing an ordinary, generally available, surplus 100-kHz crystal—such as that used in SWL and ham crystal calibrators—this simple circuit produces an accurate 100 kHz square waveform at a level suitable for reasonably good calibration of the horizontal sweep of a general-purpose scope. For example, if the scope has an 8 × 10 cm graticle, and the sweep is set to display one complete square-wave cycle per horizontal graticule division, the time-base is calibrated to 10 μsec/cm, or 100 μsec/full screen (10 divisions). If the sweep is set so that one complete cycle occupies all 10 horizontal graticule division, each division represents 1-μsec/cm.

Capacitors C1 and C2 should be placed as close as possible to the terminals of IC1. (Note. Do not substitute a "compensated" IC, such as the 741 for IC1). Zener diodes Z1 and Z2 are back-to-back with an approximate range of 1.5 to 3 volts. The output peak-to-peak signal will be twice the rating of Z1.

Ordinary 9-volt transistor radio batteries can be used for a quick-and-dirty power source.

For easiest observation on the scope, the vertical amplifier input on the scope should be set for dc. If your scope has only an ac input, there will be a decided tilt to the top and bottom of the waveform, though the displayed "frequency" (time) will be correct.

See other projects in this book for an IC1 pinout diagram.

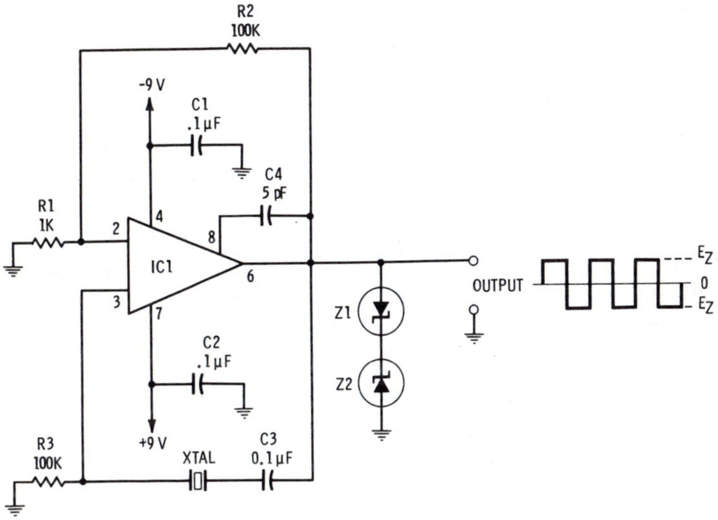

Scope calibrator

PARTS LIST

Item	Description
R1	1000-ohm, ½-watt resistor, 10%
R2, R3	100K, ½-watt resistor, 10%
C1, C2	0.1-μF Mylar capacitor
C3	0.1-μF Mylar or ceramic disc capacitor
C4	5-pF ceramic disc capacitor
IC1	Signetic NE or SE 531 operational amplifier
Z1, Z2	Zener diode (see text)
XTAL	100-kHz crystal
Miscellaneous	9-volt batteries for power supply, cabinet, etc.

34. Precision Half-Wave Rectifier

Many of the relatively low-cost chart recorders which might prove perfectly suitable for making frequency-response plots of audio equipment, or temperature and wind records, or what-have-you, are "dc only," and require either a dc voltage or current for operation.

With this precision half-wave rectifier, you can use many of the hand-size and "surplus" chart recorders for ac measurements—everything from records of the powerline variations to audio-frequency response. The only limitation is the frequency response of the integrated circuit used in this project. The specified IC will handle ultrasonic frequencies to about 100 kHz. If you are interested primarily in the audio range of 20 kHz and lower, IC1 can be replaced with a 741, in which case, capacitor C1 is eliminated.

Output control R4 is optional if your recorder requires a voltage input. If a current input is required, eliminate R4 and feed the output of IC1 directly into the recorder.

The input voltage to the precision rectifier at the location indicated as "Test Point A" should be in the range of 1 to 10 volts. If necessary, connect an amplifier ahead of Test Point A to attain the desired input voltage. In many instances, the rectifier can be connected directly to a high-fidelity tape recorder or preamplifier for audio tests, with the recorder or amplifier volume control adjusted for the correct input voltage.

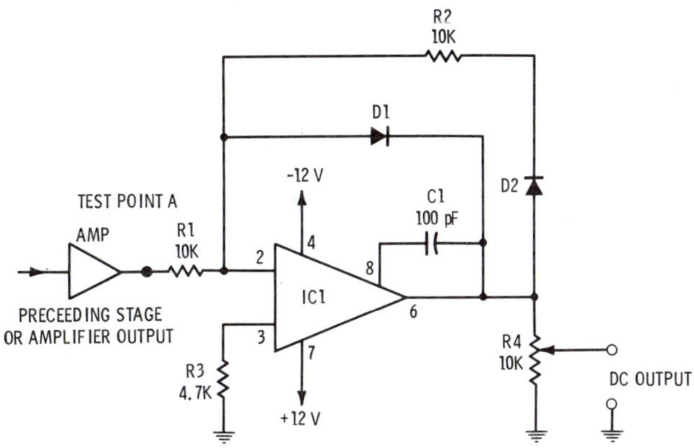

Precision half-wave rectifier

PARTS LIST

Item	Description
R1, R2	10K ½-watt resistor, 5%
R3	4.7K ½-watt resistor, 5%
R4	10K linear taper potentiometer (see text)
C1	100-pF ceramic disc capacitor (see text)
IC1	Signetics NE or SE531 operational amplifier (see text)
D1, D2	Germanium diode, 1N60 or equivalent

35. Headphone Amplifier

Several of the highest quality reel-to-reel tape recorders have monitor outputs for 600-ohm headphones, sometimes isolated from the recorder "ground." Then there are some short-wave and monitor radios with "500- or 600-ohm line level" outputs that are also suitable for headphone monitoring. Unfortunately, plugging in low-impedance hi-fi headphones (approximately 8-ohms impedance) generally loads the 500/600-ohm outputs so heavily that little, if any, sound is heard in the phones.

This headphone amplifier is especially designed for matching hi-fi phones to 500/600-ohm headphone and line-level outputs. The battery supply is 6-volts, made up of four AA, C, or D cells in a suitable holder available from most parts distributors.

Transformer T1 is a miniature 600:600 because it is the most common value to find. If you can locate a 600:10,000-ohm transformer—which is also somewhat common—it can be used if resistor R1 is eliminated. Potentiometer R2 will give a better control range if it is an audio taper, but a linear taper can be used.

Note the connections on J1 to connect the stereo phones for mono operation. If you want a stereo monitor, build two identical circuits in the same enclosure using a single 6-volt battery supply, open the jumper on jack J1 from terminals A to B and connect the second amplifier output to terminal A. For ease of volume control, ganged 10K potentiometers (often called *dual potentiometers*) should be used in a stereo version.

Capacitor C3 and resistor R5 are needed only if the circuit breaks into oscillation without them.

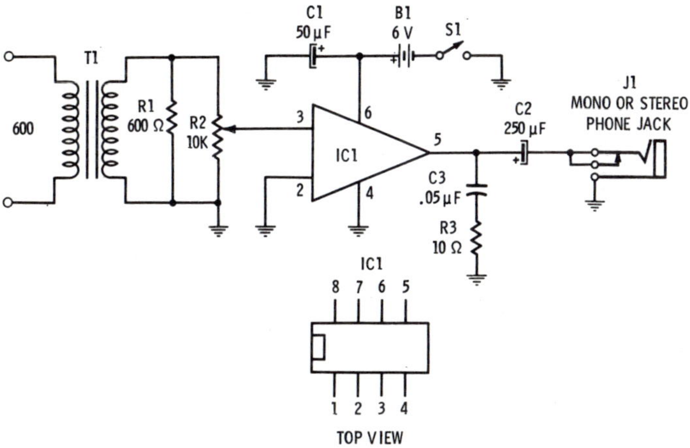

Headphone amplifier

PARTS LIST

Item	Description
R1	600 ohm, ½-watt resistor, 10%
R2	10K audio taper potentiometer
R3	10-ohm, ½-watt resistor, 10%
C1	50-μF, 6-V dc electrolytic capacitor
C2	250-μF, 6-V dc electrolytic capacitor
C3	0.05-μF ceramic disc capacitor
IC1	LM-386 audio power amplifier or equivalent
J1	Stereo or mono phone jack (as required)
B1	6-volt battery (see text)

36. LED Power Blinker

Many pieces of modern battery-powered equipment requires so little operating current that the "power-indicator lamp" often draws more than the actual operating circuit. Some equipment requires so little current it utilizes a small battery that would be drained "dry" in minutes if a pilot lamp were used, and these often have a colored flag in the power switch to indicate the power is *on*.

There is still nothing as reliable as a light to indicate power, and you will find the LED Power Blinker is an ideal "pilot lamp" for your battery-powered projects.

Powered by an alkaline AA battery this "blinky" will flash at an approximate 1-Hz rate for about six months, if left on permanently. Used intermittently to indicate that the equipment is on, the AA battery should last its shelf life of one to two years. It becomes a power indicator by simply using a double-pole switch to control both the blinker and the operating circuit, as shown in the schematic. When the operating circuit is turned on, the blinker is also turned on.

Light Emitting Diode (LED 1) can be any type and any available color, as long as its rated for at least 50-mA peak LED current. The circuit will work with LEDs of lower rating, but with so many surplus LEDs available many are bound to be "undersize" and will *pop* in this circuit; so use LEDs with a 50-mA or higher rating.

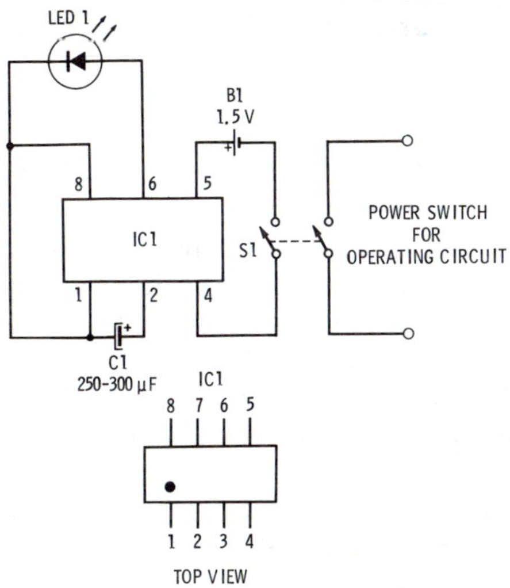

LED power blinker

PARTS LIST

Item	Description
C1	250- or 300-μF, 3-V dc electrolytic capacitor
LED 1	Light Emitting Diode rated at least 50-mA peak LED current
IC1	LM3909 or Radio Shack 276-1705 Flasher/Oscillator
B1	1.5-V battery, Type AA suggested
S1	Power switch spst (see text)

37. Portable Instrument Amplifier

Packing almost 4-watts output, this portable instrument amplifier lets you take along electronic instruments such as guitars, organs, and pianos on picnics ,outings, etc. True, electric pianos and organs require some form of basic power supply, but many models can be modified by music stores to work from a 12-volt dc supply, such as an auto battery or two series-connected 6-volt lantern batteries.

This project provides the "voice" for those instruments, and also works from the same 12-volt power source: an auto battery or a pair of lantern batteries. You can even use a 12-volt lantern battery, if you can locate a supplier (try a construction equipment or camping store).

If you are a hard-rock music enthusiast mount an in-line heat sink on the IC—any small one will do.

The entire assembly, including the speaker, should be housed in a wood cabinet that also serves as the speaker baffle. (The better the baffle, the greater the sound output and the deeper the bass.) A decent quality 6-inch or 8-inch speaker is recommended.

While provision for the batteries can be made inside the cabinet, there is less possibility of a battery getting loose in transit and causing damage if the power-supply wires are brought out for connection to external batteries.

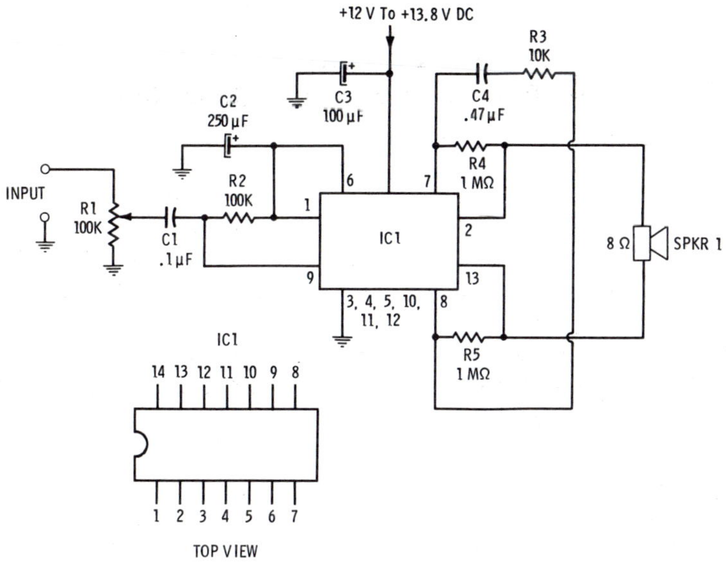

Portable instrument amplifier

PARTS LIST

Item	Description
R1	100K audio taper potentiometer
R2	100K, ½-watt resistor, 10%
R3	10K, ½-watt resistor, 10%
R4, R5	1-megohm, ½-watt resistor, 10%
C1	0.1-µF Mylar
C2	250-µF, 15-V dc electrolytic capacitor
C3	100-µF, 15-V dc electrolytic capacitor
C4	0.47-µF Mylar
IC1	LM377 dual audio amplifier
SPK 1	8-ohm speaker (see text)
Miscellaneous	Cabinet

38. Fuzzer For Electric Guitars

Soft *fuzz,* the laid-back sound of modern rock is easily built from parts that can be found in many a "hell box." And, if you do not keep "scrounged up" components, you can pick up all the parts for this project as virtually any parts distributor or surplus house.

Other than assembling the project in a metal enclosure to prevent hum and noise pickup, there are no construction problems.

The degree of fuzz, which in reality is distortion, is determined by the setting of potentiometer R1. The output level will be approximately ±0.6 volt directly at IC1 terminal No. 6. Potentiometer R4 sets the desired level into the associated guitar amplifier.

Switch S1 provides a bypass around the Fuzzer; in effect, it serves as a fuzz on-off switch.

As a general rule, the volume adjustment of the guitar should be run wide open, with the fuzz effect controlled entirely by R1. Adjust R4 so that the level from the guitar amplifier speaker is the same as when the fuzz effect is turned off.

For ease of operation, potentiometer R1 and R4 should be audio taper. If necessary, use 100K audio taper potentiometers, if a 50K audio taper is not available. But if you cannot obtain an audio taper. If necessary, use 100K audio taper potentiometers, Basically, IC1 can be any version of the popular 741 operational amplifier. See other projects in this book for the various 741 pinout arrangements.

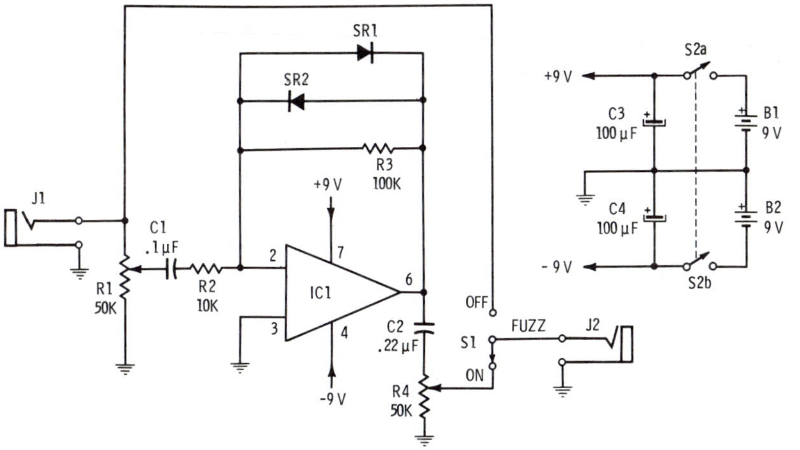

Fuzzer for electric guitars

PARTS LIST

Item	Description
R1, R4	50K audio taper potentiometer
R2	10,000-ohm, ½-watt resistor, 10%
R3	100K, ½-watt resistor, 10%
C1	0.1-µF, Mylar
C2	0.22-µF, Mylar
C3, C4	100-µF, 12-V dc, electrolytic
IC1	741 operational amplifier
SR1, SR2	Silicon rectifier or signal diode, rated 25 PIV or higher
S1	Switch, spst
S2	Switch, dpst
B1, B2	9-volt battery transistor-radio type, 2U6 or equivalent
J1, J2	Phone jack
Miscellaneous	Metal cabinet